Make:
Fun!

Make: Fun!

40 Fun Projects You Can Build!

Create Your Own Toys, Games, and Amusements

Bob Knetzger

MAKER MEDIA™
SAN FRANCISCO, CA

Make: Fun!
Create Your Own Toys, Games, and Amusements
By Bob Knetzger

Printed in Canada.

Published by Maker Media, Inc.,
1160 Battery Street East, Suite 125,
San Francisco, California 94111

Maker Media books may be purchased for educational, business, or sales promotional use. Online editions are also available for most titles (safaribooksonline.com/). For more information, contact our corporate/institutional sales department: 800-998-9938 or *corporate@oreilly.com*.

Publisher: Brian Jepson
Editor: Roger Stewart
Copy Editor: Elizabeth Campbell, Happenstance Type-O-Rama
Proofreader: Elizabeth Welch, Happenstance Type-O-Rama
Interior and Cover Designer: Maureen Forys, Happenstance Type-O-Rama
Compositor: Maureen Forys, Kate Kaminski, Happenstance Type-O-Rama
Indexer: Valerie Perry, Happenstance Type-O-Rama

April 2016: First Edition

Revision History for the First Edition
2016-04-25: First Release

See oreilly.com/catalog/errata.csp?isbn=9781680450873 for release details.

978-1-457-19412-2

Safari® Books Online

Safari Books Online is an on-demand digital library that delivers expert content in both book and video form from the world's leading authors in technology and business.

Technology professionals, software developers, web designers, and business and creative professionals use Safari Books Online as their primary resource for research, problem solving, learning, and certification training.

Safari Books Online offers a range of plans and pricing for enterprise, government, education, and individuals. Members have access to thousands of books, training videos, and prepublication manuscripts in one fully searchable database from publishers like O'Reilly Media, Prentice Hall Professional, Addison-Wesley Professional, Microsoft Press, Sams, Que, Peachpit Press, Focal Press, Cisco Press, John Wiley & Sons, Syngress, Morgan Kaufmann, IBM Redbooks, Packt, Adobe Press, FT Press, Apress, Manning, New Riders, McGraw-Hill, Jones & Bartlett, Course Technology, and hundreds more. For more information about Safari Books Online, please visit us online.

How to Contact Us

Please address comments and questions concerning this book to the publisher:

Make:
1160 Battery Street East, Suite 125
San Francisco, CA 94111
877 306-6253 (in the United States or Canada)
707-639-1355 (international or local)

Make: unites, inspires, informs, and entertains a growing community of resourceful people who undertake amazing projects in their backyards, basements, and garages. Make: celebrates your right to tweak, hack, and bend any technology to your will. The Make: audience continues to be a growing culture and community that believes in bettering ourselves, our environment, our educational system—our entire world. This is much more than an audience, it's a worldwide movement that Make is leading we call it the Maker Movement.

For more information about Make:, visit us online:

- Make: magazine makezine.com/magazine
- Maker Faire makerfaire.com
- Makezine.com makezine.com
- Maker Shed makershed.com

To comment or ask technical questions about this book, send email to bookquestions@oreilly.com.

Acknowledgments

First off, BIG thanks to my wife Deborah who in addition to being a tireless supporter and huge help to me on this book as editor and proofreader is a world-class maker herself. She's helped me on hundreds toy projects as researcher, seamstress, designer, hand model, voice-over talent, toy tester, artist, game player, sounding board, and all-around life-long collaborator. Thanks and love, Deb.

My daughter, Laura, pitched in to test, make, and help photograph several of the projects in the book. She's also a fine cartoonist and drew the fun little MakeyBot mascot illustrations.

My son, Reed, helped me build and test the strip heater project, although he had an ulterior motive in getting a custom-built part for his sports car. (I encourage all parents to make stuff with their kids—you'll learn a lot!)

Some of these projects are spin-offs of things I had created with my toy-design business partner Rick Gurolnick. Thanks, Rick, for making NeoToy with me. Your cleverness, business sense, and humor are inspiring.

Mark Frauenfelder asked me to write for *Make:* magazine, starting way back on volume 11. He was a writer's dream editor-in-chief; full of encouragement for my DIY projects, along with enthusiastic and creative guidance for my editorial articles, some of which are contained here. Makers: it takes one to know one. Thanks, Mark.

This book wouldn't have happened without the dedicated work of all the people at Maker Media. Gratitude is due to my book editor and fellow pop culture fan Roger Stewart for championing a fresh new design for this book, along with editor Brian Jepson, wordsmith Elizabeth Campbell, and layout whiz Maureen Forys. My thanks also to *Make:* magazine editor Keith Hammond, who keeps it all going smoothly. Goli Mohammadi and Katie Walker Wilson helped make things look great and Jeffery Braverman, along with Gregory Hayes, took some nifty photos! And of course thanks to maker-in-chief, Dale Dougherty. Together we've had more than few "issues" at *Make:* magazine—all of them good!

A special tip of my safety goggles to *Make:* readers who generously shared their creations and photos. Thanks to David Corina, Troy Fischer, Pat and Francis Fullam, Derek Gable, Doug Stith, Kurt Gulatieri, Jack Knetzger, Greg Lehman, Bob Marino, Richard and Julie Shiffman, Steve Wacker, John Diehl, and the Girl Scouts Heart of the Hudson.

Thanks to these esteemed toy industry professionals for help with research on the Marvin Glass article: George Gomez, Eddy Goldfarb, Richard Levy, and Pauline Camberlis.

—BOB KNETZGER

About the Author

Bob Knetzger has been making fun stuff all his life—first as a kid with his trusty Vac•U•Form and Thingmaker toys, then as a professional toy designer at Mattel, and for the last 35 years as an independent toy inventor. Bob's designs for toys and games have appeared on television, in hundreds of millions of cereal boxes, in the pages of *Make:* magazine, and *now* in your very hands!

Contents

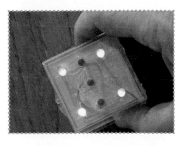

Chapter 5   Casting and Molding. 137

Chapter 6   Holiday Projects 161

Chapter 7   Working with Metal. 183

Appendix   Bonus Material 213

Mad Monster Candy Snatch game
face label

Foreword

I feel very lucky to have been a little kid in the 1960s. The toys were fantastic!

Here are the ones that made a big impression on me:

- The *View-Master* and the reels of 3D artist Florence Thomas's fairytale land dioramas, which gave me the feeling that I was peering into a magical universe just behind the stereo lenses of the plastic binocular-like viewing device. I treasured my *View-Master*, and any friend of mine who pulled down on the lever and let it snap back was forbidden from using it.

- The *Air Blaster*, made of a similar ominous black plastic as the *View-Master*. It looked like a pistol with a large funnel attached to the barrel. It shot air vortices that could knock over paper targets or blow out candles from across the room. The burst of air was so powerful that I'm sure it could have popped an eardrum if you placed it over someone's ear and pulled the trigger. I never did that, but I did like to fill it with juniper berries and shoot them against our living room window, which made a lot of noise.

- *Creepy Crawlers*, which was one of the *Thingmaker* line of toys. This kit came with an A/C powered hotplate; metal molds of worms, spiders, snakes, and other scary bugs; and several squeeze bottles of a delightful-smelling liquid substance called Plastigoop that cured when exposed to heat. This kit kept my sister and me occupied for hours. It was fun to plan which colors of Plastigoop to use, and it was fun to use the tongs to set the mold on the hot plate, and it was fun to quench the mold in the cold water bath, and it was fun to peel the crawlers out of the molds (especially when you succeeded in doing it without pulling a spider's leg off by mistake).

- The *Show'N Tell*, which played records and projected film strips on a simulated TV screen. I was thrilled by the "Dinosaurs" Picturesound program, which played "Night on Bald Mountain" when they talked about the Tyrannosaurus Rex.

The thing I liked about these toys was that they occupied a sweet spot on a spectrum between complete lockdown and total open-endedness. They had a specific function that allowed for a wide variety of amusement. They offered just the right amount of guidance for a little kid but encouraged experimentation (or, in the case of the *View-Master* and the *Show'N Tell*, offered lots of different stories you could select "on demand" by grabbing a disc or reel from your toybox).

I like Bob Knetzger's toy inventions for the same reason. Bob grew up playing with the same kinds of toys as I did, and so he shares a love for toys that do cool things and offer a type of guided creativity that is too often lacking in many of today's toys. Bob has had decades of experience inventing toys, and he understands them on many levels. He is a talented artist and designer, and he understands mechanisms, electronics, and materials. This is a killer combination for a toy designer, so it isn't surprising that major toy companies compete for his inventions.

I don't think Bob designs toys that he believes will appeal to kids. He designs toys that appeal to him. And because of this, his toys have a sense of humor and playfulness that shine through. He loves toys and everything that goes into making great ones. Bob is very generous to have written this book, which not only has specific step-by-step instructions for making excellent toys, but also includes his hard-won insider information about the art of toy invention.

I've read a lot of books about how to make toys, but this is the first one that gets to the heart of the innovation and creativity that goes into being a toy inventor.

—MARK FRAUENFELDER

Introduction

In my many years of designing and inventing toys, I've learned the most just by making. I was always eager to get started on a given project and make my own discoveries. If I hit a setback then I'd get help as needed. I hope you'll take the same approach with the material in this book. Let the designs and plans serve as a starting point and an inspiration to what *you* want to make.

A couple of the projects are what I call *meta-projects*: projects that you first build and then use to make *other* projects. Make the Kitchen Floor Vacuum Former, and then use it to mold plastic parts for other projects. Same for the Foam Cutter and E-Z-Make Oven: use them to cut foam and cook materials to make what you want.

Many of these projects, tips, and ideas are spin-offs and examples from my work as a professional toy and game designer. Some projects are super simple; some are more involved. No expensive or highly technical tools are needed, just basic tools found in any garage or hobby bench. No specialized skills are required either.

I've included some clip-and-use pages in the Appendix to give you a head start on many of the projects. Look there for some other fun bonus stuff, too.

You'll find QR codes throughout the book. Scan them with your smartphone to see short demos of the projects in action. (Can't wait? There's a tiny flipbook animation you can view right now—just use your thumb to flip the pages and see the spinning turbine in action from the Desktop Foundry project in Chapter 7.)

See more Make: FUN! goodies online at makerfunbook.com: printable PDFs, videos, and more. Also online is a *bonus* chapter, "Industrial Design for Makers." Look over my shoulder as I use industrial design processes and techniques to remake a popular project from *Make:* magazine. See how I transform this:

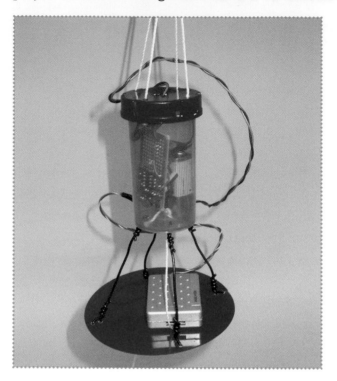

. . . into this:

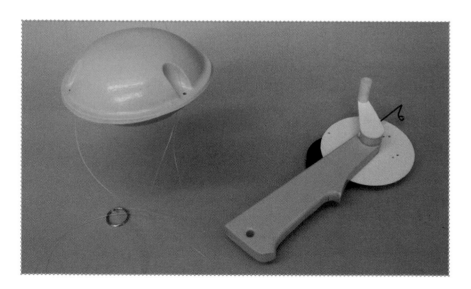

Ask me a question or send me an e-mail with pictures and videos of your own projects and I'll share them there for everyone to see, if you like.

You might make some of these projects with a kid, for a kid (see "Play Safe!" below), or just for yourself. I hope you'll have as much FUN reading about and making the things in this book as I had creating and writing it.

—BOB KNETZGER

 **→ Play Safe! ←**

The fine print: Technology, the laws, and limitations imposed by manufacturers and content owners are constantly changing. Thus, some of the projects described may not work, may be inconsistent with current laws or user agreements, or may damage or adversely affect some equipment. Your safety is your own responsibility, including proper use of equipment and safety gear, and determining whether you have adequate skill and experience. Power tools, electricity, and other resources used for these projects are dangerous unless used properly and with adequate precautions, including the use of proper safety gear. In order to show the project steps more clearly, some illustrative photos do not depict safety precautions or equipment. These projects are not for use by unattended children. Use of instructions is at your own risk. Maker Media Inc. and Robert Knetzger disclaim all responsibility for any resulting damage, injury, or expense. It is your responsibility to make sure that your activities comply with applicable laws, including copyright.

Chapter 1

Toy Inventing and Kit Bashing

*H*is innovative products changed the world, yet he was never satisfied and was always working on the next thing. He was a flamboyant showman, and delighted in cleverly presenting his latest top-secret ideas. He earned the fierce loyalty of his employees, although he could be exceedingly difficult. He was a genius when it came to design, but he didn't actually invent his most famous products. His personal life was complicated and he was moody, but his products brought joy to millions. And after his untimely death, his most famous products live on and are as popular as ever. Although this might also describe Apple cofounder Steve Jobs, the complicated creator profiled here is toy innovator Marvin Glass.

Try me!
Use your thumb to quickly flip through the pages (back-to-front).
See the spinning turbine from Chapter 7 in action!

Marvin Glass,
Titan of Toy Invention

Baby Boomers may not have known his name, but Marvin Glass's amazing toys and games and their TV-promoted jingles were unforgettable: *Rock 'Em Sock 'Em Robots* ("His block is knocked off?!")…"Open the door to your (*ahhhh!*) *Mystery Date!*"…"*Operation!* (*buzzzz!*)." Some Marvin Glass toys are distant memories, but many remain popular and still sell today. Glass's greatest triumph was not any one particular toy or game: Glass perfected the *business* part of the toy invention business.

Born to German immigrants in 1914, Marvin grew up near Chicago. During an unhappy childhood he created his own toys from cardboard and wood: a toy dog, swords and shields, and a climb-inside toy tank. Foreshadowing a lifelong pattern, his toy creations made his friends happy, yet he remained lonely.

He designed animated store window displays after college and sold an invention for a toy theater to a manufacturer for $500. When he learned the company made much more from his idea ($30,000!), Glass realized what he must do in the future: *license* his designs and earn a royalty on each one sold.

An inventor named Eddy Goldfarb showed his invention, *Yakity-Yak Talking Teeth*, to Glass. They struck a deal: Goldfarb did the inventing and Glass did the selling. Glass sold the design to novelty maker H. Fishlove and Company of Chicago, and

the comical wind-up chompers became a hit in 1950. Glass promoted other Goldfarb creations—and himself in the process. Goldfarb told me: "Marvin was a great, *great* salesman. He should have been in the movie business!" Goldfarb left Glass to create hundreds of toys on his own, like Mattel's *Vac•U•Form* and Schaper *Stomper* mini-motorized cars. Goldfarb was the true pioneer of the *invention* part of the toy invention business.

Glass created a lavish, top-secret toy studio in Chicago worthy of Willy Wonka. Instead of Oompa-Loompas he hired talented designers, inventors, sculptors, and model makers. Lathes and mills were painted in bright colors. Chagall paintings and Remington sculptures filled the waiting rooms at Marvin Glass and Associates (MGA). Every new toy or game idea was kept absolutely hush-hush. Soundproofed windows and triple-keyed locks prevented corporate espionage, all monitored by closed-circuit surveillance cameras. Work-in-progress prototypes were locked up in a safe every night. Was Glass's paranoia justified? Probably not, but the mystique of such dramatic flourishes only added to the perception of visiting toy company executives: these ideas must really be something! According to writer and toy inventor Richard C. Levy, salesmanship is "not what you have—it's what they *think* you have."

And what ideas they were! MGA's new toy concepts had clever features that often created all-new categories of toy and games.

Children's board games had been a sleepy product category with such staid offerings as

Candy Land and *Chutes and Ladders*. "Bored" game was right. MGA changed all that with their innovative "skill and action" games. *Rock 'Em Sock 'Em Robots* took its inspiration from an arcade boxing game, but was miniaturized for tabletop action and added a telegenic gimmick: landing a solid push-button punch on the robot's jaw popped his head up with a revving *buzzzzzzzz!* sound.

MGA's game-plus-toy hybrids (*gamoys* in the toy biz lingo) transformed the flat game board into three dimensions. *Mouse Trap* was inspired by Rube Goldberg's comic. It was masterfully rendered with styrene gears, ramps and chutes, springs, rubber bands, and marbles. The game play wasn't much more than ritualized assembling of the toy contraption. No matter. The real fun was activating the chain reaction of mechanical gizmos that ultimately dropped a cage over a mouse token, ending the game. A huge hit in 1963, it sold virtually unchanged for over 40 years. MGA attempted to clone *Mouse Trap*'s success with two sequels, *Crazy Clock Game* and *Fish Bait Game*, but the timing was wrong and kids didn't bite. That's the fickle toy biz!

Robot Commando was a great example of a Marvin Glass toy. The exciting 1960 TV commercial showed a giant, motorized robot that listens and obeys. Say, "Forward!" into the mouthpiece control and the robot goes forward. Say, "Fire!" and he shoots missiles from his domed head or flings balls from his spinning arms. Wow! What boy wouldn't want a robot that responds to verbal commands? But it's not what you think. MGA's clever design featured a wired remote control with a Bowden cable (similar to the gearshift control on a 10-speed bicycle). When you twisted the knob to move the pointer from Forward to Shoot, a stiff cable pushed or pulled to mechanically shift gears inside the robot. At the same time, the vibrations of your voice sympathetically jiggled a metal contact, which connected the battery power and energized the motor. You could say, "Fire!" but if the knob pointed to Turn Left the robot turned left. Dramatic, successful, and artfully deceptive, *Robot Commando* was not unlike Marvin Glass himself.

Toy manufacturers like Ideal, Hasbro, and Louis Marx lined up to get a peek at MGA's latest inventions. The string of hit toys seemed endless: take-apart, put-together robot *Mr. Machine* (1960), slap-action card game *Hands Down* (1964), toss-the-hot-potato game *Time Bomb* (1964), glowing, colorful peg-picture maker *Lite Brite* (1967), free-wheeling *Evel Kneivel Stunt Cycle* (1973). You couldn't turn on a TV, open a Sears Christmas catalog, or go into a kid's bedroom without seeing an MGA design.

Although Marvin Glass's name and logo was on the box, others created the inventions inside. Inventor Burt Meyer developed *Lite Brite* and *Rock 'Em Sock 'Em Robots*, and, along with Leo Kripak, created *Mr. Machine*. The idea for the game *Operation* originally came from John Spinello, who accepted Glass's offer of $500 and an unfulfilled promise of a future job. Gordon Barlow and Meyer co-invented *Mouse Trap*. Carl Ayala came up with *Whoops* fake vomit despite his boss's initial disapproval.

It was a high-stakes, high-pressure business. His secretary of many years, Pauline Camberis, says, "I loved working for Marvin. He was very difficult at times, but he was that way with most of his employees. Marvin was also very generous."

All this success enabled Marvin to indulge in a luxurious lifestyle. He had a chauffeur to drive his clients around and a personal chef to cook them gourmet meals. His remodeled Evanston, Illinois, carriage house residence appeared in a 1970 Playboy magazine photo layout, complete with a cocktail table with built-in hi-fi controls (swanky!) and party guests in the sauna and whirlpool (swingin'!). And yet, Glass remained disappointed with himself. He was famously profiled in *The Saturday Evening Post* as the "Troubled King of Toys." His first ex-wife said, "I think he was born to be dissatisfied. That's why he can keep on creating."

Marvin Glass died in 1974 at age 59. His business partners kept the firm going, inventing and licensing toys and games, but MGA finally closed in 1988. Many of Glass's partners and employees have spun off to work on their own in the toy invention business, Marvin Glass's greatest creation.

→Hits (and Flops)←

Glass revolutionized the game business with 3D game/toy hybrids, most famously *Mouse Trap*, still made today. Here are the MGA hits...and flops!

MOUSE TRAP (Ideal, 1963)—A HIT!
This iconic tabletop game has inspired this *giant* version, *Life Size Mouse Trap*, seen at Maker Faire. It uses actual bowling balls instead of marbles!

CRAZY CLOCK GAME (Ideal Toys, 1964)—A FLOP!

A similar concept to *Mouse Trap*, it was more toy than game, with a series of contraptions that create a chain reaction, the final stage being a sleeping figure flying up out of bed. The individual mechanisms include levers and swivels, with stored potential energy in the form of elevated marbles and energized springs. Like *Mouse Trap* before it, *Crazy Clock Game*'s design includes various escapement mechanisms to s-l-o-w down the action, like a marble rolling down a crooked staircase, or a spinning wheel that slowly rolls across a track. The design created humorous drama and tension but failed to create major sales.

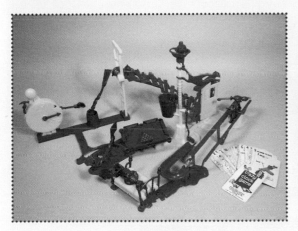

FISH BAIT GAME (Ideal Toys, 1965)—A FLOP!

Maybe the goal was to make a cheaper, cost-reduced version of the prior two items. But with skimpy action (bobbing pelicans) and less witty mechanisms (a rubber band–snapping finale), *Fish Bait Game* was the weakest entry in the MGA contraption-game trilogy. It's a rare collector's item today.

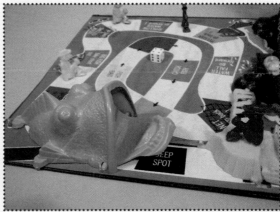

TIGER ISLAND (Ideal Toys, 1966)—**A̶ FLOP!**

A reaction-time game with a spinning tiger and marble-tossing castaways—stranded in the toy aisle! A truly bizarre game: the center tiger swivels around and, if you were unlucky, would randomly stop in front of your castaway figure. You have only a split second to launch a marble into the tiger's mouth and jam the mechanism. If you miss or are too late, the tiger klonks you on the head with his oversized club— you lose. The first player to collect enough rods to complete their escape raft is the winner.

PERILS OF PAULINE (Marx Toys, 1964)—**A̶ FLOP!**
Sculpted and painted 3D game components couldn't rescue a boring game.

Mystery Date (Milton Bradley, 1965)—**A HIT!**

Who waits behind the door? Will you get a dreamboat? Or a dud?

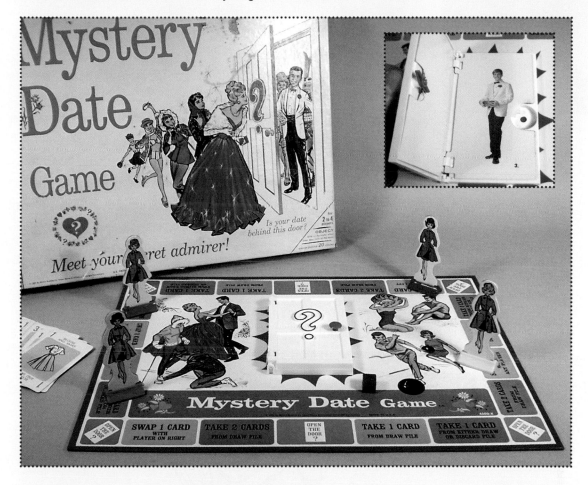

How did it work? Behind the door were five different picture panels, like pages in a book. Each panel had a differently shaped edge. Note the blue plastic paddle on the back of the door. This was connected to the doorknob on the front. You'd spin the knob, then open the door. The lobes on the paddle would line up with just one of the panel edges and open the "book" to that "page." Spin the knob for a different random date. It was a clever low-cost randomizer gimmick with a mystery outcome every time!

Fun Fact: Marvin Glass shrewdly negotiated to have his company's logo printed right on the game box—hit or flop.

Steampunk Safety Goggles

Toy inventors need to quickly turn their ideas into prototypes in order to test a new concept before investing a great deal of time. Rather than draw up detailed plans and laboriously machine custom parts, they just adapt parts and mechanisms from existing items to slap together a rough-and-ready breadboard model. You learn more about what you need to do faster just by doing it!

If a model maker needs to build a visual "looks-like" model with intricate details, like a space vehicle, they often *kit bash* parts from a completely different model: the rear-end differential from a hot-rod car model becomes the pressure tank and thruster engine on a rocket ship!

Try this easy project! There's no wrong way to build it. Whatever *you* like, goes. With a little bit of scrounging and a touch of paint you can transform ordinary safety goggles into something dangerously cool: a steampunk-themed pair of goggles complete with sci-fi gizmo gears along with Victorian-era gold work and turn-of-the-century tech graphics.

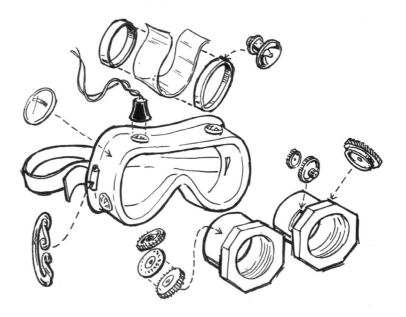

You'll need some simple hand tools—saw, drill, sandpaper—plus a hot-glue gun and black and gold spray paint. Gather up some interesting bits and bobs and, of course, a pair of safety goggles. I used the big, one-piece design that fits over my glasses. They have big surfaces on the sides and top for attaching steampunky stuff.

No need to follow a detailed plan; create your own design. Just drill or cut the parts, as needed, and hold them up to the goggles. Try out how they look in various positions, then hot glue them in place.

I turned these safety goggles into a fun, fake, night-vision goggle costume. A pair of eight-sided, threaded, plastic plumbing connectors became "night vision" scope tubes. I hot-glued a handful of plastic gears from a broken toy on top, together with a number dial from a disposable camera for some interesting scientific-looking details. The bulb and reflector scrounged from a penlight, along with coiled wires and bogus connections, complete the night-vision look. I added a cylindrical storage compartment by rolling a piece of thin plastic inside some plastic rings and hot-gluing that on top in the center. The definitely unsafe green cellophane lenses are optional.

For a more Victorian feel, I included some scrolled chrome buckles for the strap and Photoshopped a few labels to make dials and gauges, as well as an appropriate nameplate. (Look in the appendix for some clip-and-use graphics to use for your own project.) To finish it off, I sprayed a little flat black paint here and some gold paint there. Voilà!

MAKER TIP

Kit bashing is a term model makers use to describe taking parts from different model kits and using them to fabricate something else quickly. No need to hand-make custom detailed components from scratch when there are probably already plenty of interesting shapes in your junk drawer or garbage can. Broken appliances, old toys, or obsolete electronic items contain lots of cool-looking and useful parts, so before you toss them out, get in the habit of harvesting these bits of treasure from the trash: gears, knobs and dial faces, pulleys, belts, and more.

Even salvaged screws and fasteners can come in handy. Plastic appliances and toys are often held together with special thread-cutting screws. They have a tapered shape, sharp threads, and a functional small notch in the tip that acts as a relief and screws better into plastic or wood.

I save all my scavenged parts in sorted boxes and mini drawers.

Footstep Sand Stampers

Here's an easy project for summer fun: hack a pair of flip-flops so they make designs in the sand as you walk. Just add a bit of Sugru® to your sandals and start stamping!

First, clean the bottom of your flip-flops thoroughly with some vinegar and water to get rid of any grease or loose dirt. Dry off the soles completely so the Sugru will stick and stay on. Open the pack and knead the Sugru until it's pliant and soft, then form the shapes you want.

Make shapes by first rolling out a small snake, then press it onto the sole of the flip-flop. Spread the bottom out and press down firmly for good adhesion. Don't make tiny details (you won't really see them in the sand)—think simple and *bold* shapes! Make letters, numbers, emoticons, symbols, and designs, but whatever you make, remember to make it *backward*, like on a rubber stamp!

Press down all around the edges of the Sugru. Form a triangular cross-section to maximize adhesion and eliminate any undercuts for best results. Let the Sugru cure for 24 hours, then go sand stamping!

Works best in moist sand. I got great results, even with a tired, old pair of flip-flops and some almost-expired Sugru.

What design will *you* make? Happy sand stamping!

  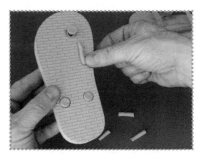

  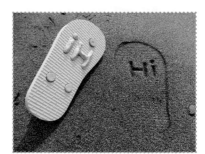

MAKER TIP

This is another project that uses Sugru, the amazing material that's fun and easy to use. You can use it to mend or hack just about anything. It starts off soft and pliable so you can shape it with your hands and press it to any clean, dry surface. After 24 hours it cures into a soft but tough rubber-like material that *really* sticks—even to glass! It's often used to repair cords, mend broken housings, or add soft safety bumpers to sharp corners or edges. I've used it to repair a torn vinyl refrigerator door seal. I also use Sugru as self-setting rubber for quickly molding parts (see "Desktop Foundry"). Sugru comes in a fun range of colors and you can even mix custom colors to create any hue. It's good to have some on hand for your next house-hold emergency. If you can't find this English elastomer locally you can get some online at the Maker Shed.

Fun fact: The name Sugru derives from the Irish word *súgradh*, meaning "play." (Disclaimer: I do not work for Sugru—but Sugru works for me!)

Hummingbird Feeder Hack

This quickie project is a good example of kit bashing: improvising a new use for an old toy. Interestingly, after this piece was first published, I saw an updated hummingbird feeder design at the pet store. It has a similar moat already molded right into the plastic cap!

Our backyard hummingbirds were *not* happy! A continuous trail of hungry ants was climbing right into our hummingbird feeder, fouling the sugar water and keeping the hummingbirds away. The ants were not deterred by folk remedies like bay leaf barriers, and pesticides were out of the question. Ugh—what to do?

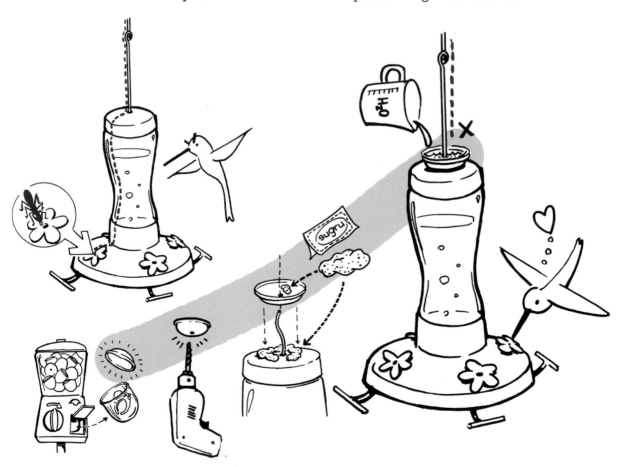

My bin of old toy parts yielded the answer. I used the cap from a vending machine gashapon toy capsule to make a water barrier. To make one of your own, drill a small hole in the cap. Attach the inverted cap onto top of the feeder with a bit of Sugru or bathroom caulk. Add some more all around the center hole on top, as well. When set, it holds the cap on and seals up the center hole where the hanging wire goes through. Reassemble the feeder, hang it back up, and fill the cup with water. The cap makes a tiny water moat that the ants refuse to cross. The hummingbirds are now back to enjoying the ant-free feeder—sweet!

Chapter 2

Easy Electronics

If you're new to making electronic projects, you'll need to build up some basic technical skills like wiring, soldering, and identifying electronic components. (If you are starting from scratch, check out Charles Platt's excellent guide, *Make: Electronics*.) No worries: even a beginner can build the following easy electronic toy and game projects. You'll learn how to use the same electronic sound circuit to make two different projects: a retro 1980s music toy *and* conductive ink game cards. But first up: projects that show how to quickly hack an existing toy to add a new "fun-ction."

Noninvasive Hacked 'Bot

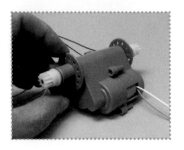

When making new toy prototypes, I often harvest old broken toys to reuse the parts such as motors, gear trains, or radio control transmitters and receivers. No need to reinvent the wheel—literally!

SmartLab Toys' ReCon is a programmable tank-treaded robot that's full of goodies. To play, write a program of step-by-step instructions for the robot to do; for example, "Go forward 48 inches . . . turn left 90 degrees . . . STOP! . . . play a sound . . . go backward 24 inches . . ." Then press Go and the program instructs the robot how to execute the mission. Open ReCon up and you'll find lots of cool stuff inside, including a nifty dual-motor drive module with built-in optical wheel counters.

Rather than take it apart, here's a *noninvasive* hack that takes advantage of ReCon's features, while adding a fun new function, to make a Root Beer Pong 'Bot. You'll need just a few small pieces of thin metal (I used brass), a light bulb, a momentary contact single-pole, single-throw micro switch, and some wire.

I wanted to add a sensor to control ReCon in real time. Fortunately the toy's nonvolatile memory retains your program, even when the batteries are removed. I used this feature to add a kill switch, something that temporarily interrupts the battery power to the toy.

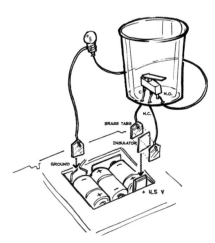

I made a thin, double-sided contact that slips in between the batteries and one of ReCon's battery contacts. Each side has a piece of brass soldered to wire with a nonconductive piece of plastic double-sided tape in between. One brass tab touches the battery to route the power to the SPST switch. Normally, the power is sent through the switch right back to the toy through the second brass tab.

But if the switch is activated, the power is disconnected from the toy and instead goes only to the light bulb. ReCon stops in his tracks and the bulb lights up—that's a "kill."

Next, I created a simple program that turns ReCon into a moving target game. As it follows a programmed path across the floor, it also plays a series of sound messages announcing an ever-decreasing point jackpot. Try to toss the ball into the cup, throw the switch, and stop ReCon before it counts down to zero. Whatever point value you last heard is your score!

If you want to try programming this game into your own ReCon, look in the appendix for a listing of my program.

A close-up of the battery compartment shows the two wires with a thin insulator between them at the lower right. The red wire steals the power from the batteries, sends it to the outboard switch, which sends it either right back via the green wire or to the external bulb.

Throw the ball in the ReCon's cup before it gets away!

The weight of the ball triggers the switch, stopping ReCon and illuminating the "kill" light.

Maker Tip

And when you're done with this noninvasive hack, just slip the wires out from the battery compartment and your toy is back to original factory condition! You can try this type of power-robbing circuit hack with lots of toys and gadgets.

➤Hack in a Hat⬅

Here's how I hacked an iPod to sell an idea.

When pitching a new idea to a toy company, I need to demonstrate the fun and excitement quickly without spending too much time on an idea that might not pan out. That was the case when I presented this "game-in-a-hat" idea: a preschooler-sized hat uses a motion sensor and sound circuit to play a simple game. As the hat plays a song, the kid dances along and the motion sensor in the hat monitors the kid's actions. When the music stops, the kid must freeze. But if he moves—*RAZZZZZ!*—he's out. Listening and dancing fun!

But how could I make a compelling presentation without having to completely build and program a working prototype? I made a hat and added a speaker and push-button SPST switch, both wired to an iPod through a wired control. My SPST switch was twinned to the circuit board traces for the Next Track button on a remote, and the speaker was connected to the Audio Out earphone jack. Every time I pressed the push button on the hat, the iPod would skip ahead to the next audio track, as if you had pressed the Next Track button on the player. Simple!

I recorded multiple audio tracks (including some silent "spacer" tracks) and made a cleverly sequenced playlist that simulated how the real toy would play the game. The voices and music played through the hat's speaker while I surreptitiously pressed the button to trigger the next audio track as-needed to demonstrate the game. As

long as I didn't deviate from my canned routine, the musical hat demoed exactly like the real thing. I wore the hat and danced or froze throughout my demonstration.

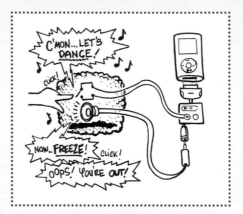

The toy company loved it and bought my idea! Here's how it came out:

What cool gadget can *you* hack from an iPod sound player, or some other device around your house?

Scan to see the TV commercial for this "game-in-a-hat" toy!

Super-Cheap Electronic Die Challenge

Suppose you needed to build an electronic die for a board game. Sure, you could use discrete components to build a clock circuit that sends a continuous signal into a decade counter and binary-coded decimal to seven-segment decoder, for example. Or you could write some random number generator code for an Arduino and use discrete LEDs. That's obvious and has been done before. But professional toy inventors have to be *crazy* clever and get the same effect but for an insanely low, dirt-cheap cost. Can you do it?

Here's one approach that costs next to nothing: LEDs wired up with some cleverly designed single-pole, double-throw switches. Each switch is made of three parallel wires and a metal ball in a cage. When you shake the cage, the ball randomly lands on either one pair of wires or the other pair. Only one pair of wires completes the circuit, lighting the LED.

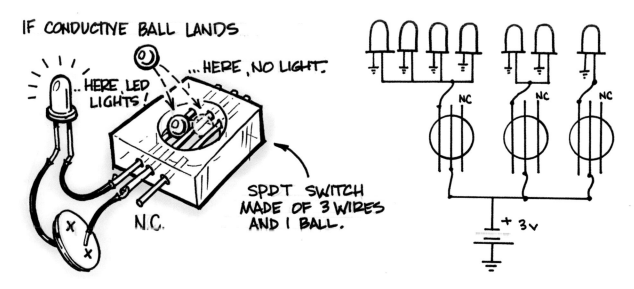

IF CONDUCTIVE BALL LANDS

..HERE, LED LIGHTS!

...HERE, NO LIGHT.

N.C.

SPDT SWITCH MADE OF 3 WIRES AND 1 BALL.

NC   NC   NC

+ 3v

Here's how I made an electronic die: Wire up four LEDs to one switch, two LEDs to the next switch, and a single LED to the last switch. Shake them all together and you'll get a random number of LEDs to light up, from zero to seven. For games that need a number from one to six, just "roll again" if you get zero or seven lit LEDs.

Here's your challenge: can you rewire the switches so that you'll get *only* numbers one through six? That is to say, no zeros and no sevens. Here's a blank diagram. Go ahead and draw your wiring here, but use a pencil: you may need to erase as you go. When you're done, go to the appendix to see my solution!

Maker Tip

Hint: The new version may not be a "fair" die, but its average roll yield *will* be the same as a real die.

This challenge is more of a thought experiment, but if you want to try actually building the die, the following are step-by step instructions.

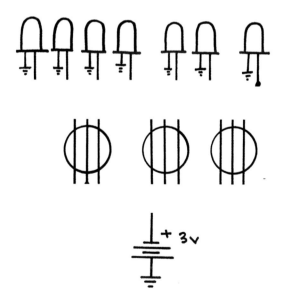

MAKE:

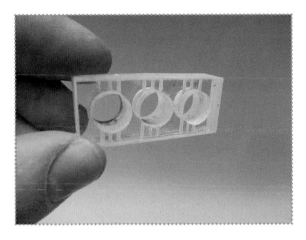

1. Make a ball cage from acrylic by drilling three large through-holes (three times the diameter of your conductive ball) and then three small holes at 90 degrees to each large hole.

2. Make a small square panel to just fit inside the box, and drill holes for the seven LEDs.

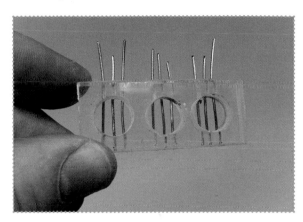

3. Strip the heavy gauge solid core wire, cut into pieces, and thread into the small holes. Superglue the ends to tack it all in place.

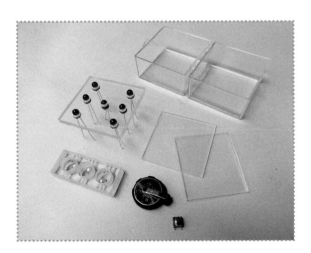

Materials

7 LEDs
Battery holder and battery
3 small metal balls (test with a continuity checker to make sure they are good conductors)
Various thicknesses of acrylic
Clear styrene snap cube
Solid and stranded wire
Momentary SPST switch for an on/off power switch (optional)

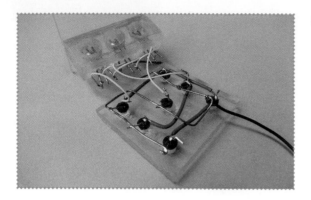

4. Solvent-bond the ball switches to another small vertical piece of acrylic so that the wires are held horizontally. Cover with a clear piece of plastic on top to trap the balls inside. Wire up the switches and LEDs.

TEST:

5. Wire up the battery and optional on/off switch and test the circuit by making a truth table of all inputs and noted outputs. LEDs one through six should light up; no zero and no seven.

6. Assemble and carefully solvent-bond the panels inside the cube (cut a hole in the back for the on/off switch).

ROLL:

7. Your electromechanical die is complete.

8. For a more finished look, paint the cube white. Leave the top clear and then add a piece of frosted mylar on top for a backlit look.

Talking Booby Trap

Having trouble with people snatching your top-secret stuff? Need help getting some privacy? Here's a sneaky MacGyver-y gizmo you can make to keep those snoops away! It's a talking booby trap: record your personalized message or sound effect and then hide it in a strategic place. If it's disturbed, the intruder will hear your message telling him or her to get lost!

Materials

Recording module

Clothespin (the kind with a metal spring)

Plastic bottle cap from a milk jug

Zip ties

Hook-up wire

Rubber band

Double-back foam tape

Aluminum duct tape

Brass strip

9V battery

(Not shown: two small wood screws)

Tools

Soldering iron and solder

Wire cutters/strippers

Drill and small bits

Scissors or hobby knife

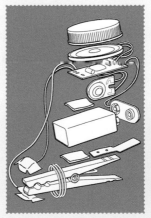

This design uses a sound recording/playback unit available from RadioShack (part 276-1323). It's a prewired module, complete with audio board, speaker, and controls, that will record up to 20 seconds of sound in nonvolatile memory. As of this writing, you can still get this module in remaining RadioShack stores or online, but if you can't get it, you can use any record-and-play toy.

You can also hack the guts of a sound recording/playback greeting card.

1. Disassemble the clothespin—the wire spring is too stiff. Reassemble the pin as shown so the spring is used only as a fulcrum pivot. Wrap a rubber band around the jaws of the clip a couple of times. Slide the rubber band closer or farther away from the fulcrum to adjust the tension. The clip should open easily, but still close together all the way.

   Add an extension to the top leg of the clothespin. Bend a small strip of stiff brass so that it lies flat when the clothespin is held open. Drill 2 small holes in the brass and attach it to the clothespin with screws.

2. Wrap the jaws of the clothespin with the aluminum tape. Poke a small hole in the aluminum and attach a short wire to each jaw. Twist the wire and crimp the tape over firmly to make a good electrical connection.

3. Install the 9V battery and test the circuit: press and hold the Record button (the red LED goes on) and speak loudly into the speaker. When you're done, release the record button, then press the gray button on the PC board to hear your recording.

4. You can improve the sound of the naked speaker significantly by adding a resonant chamber. I used a plastic bottle cap from a gallon milk jug; it's just the right diameter. Superglue it to the front of the speaker.

5. Now modify the circuit to wire up the clothespin. Look for the gray rubber-domed Play switch on the PC board. Bend the three metal tabs on the back and remove the button.

6. Feed a wire from the clothespin jaw through the tab hole and over the edge of one of the traces. Carefully solder the wire to the trace—don't short out the traces! Do the same thing for the other jaw wire and solder it to the remaining trace.

7. Now the clothespin will act as a Play switch: when the jaws touch, the sound plays. Try it! To prevent the sound from playing, place a slip of paper as an insulator between the jaws.

8. To finish, stick the battery to the clothespin with double-sided foam tape, and then stick on the PC board. The speaker is foam-taped on top. Use cable ties to cinch everything together and tuck in the microphone and Record switch wires to neaten it all up.

9. Use your Talking Booby Trap lots of ways:

- Place a diary, journal, or any object on the brass tab. The weight of the object keeps the clothespin open and armed. Camouflage the trap by placing something in front. If anyone lifts the book—WHOOP!—the alarm goes off!

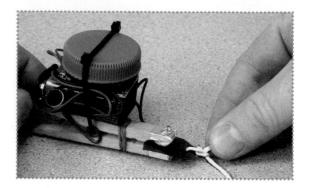

- Open the Talking Booby Trap and slip it under a closed door. If anybody opens the door, you'll know it. Record a scolding "Bad dog!" message in your own voice to keep pets in their place while you're away! You can also arm a drawer or sliding door.

- Use the Talking Booby Trap to shame your lunch-lifting coworkers. Hide it behind your food items inside the fridge at work. If anybody touches your lunch, out blasts "KEEP YOUR HANDS OFF MY LUNCH!" Now everybody will know who the secret food-stealer is!

- Tie a string to a small piece of paper and slip it in between the clothespin jaws. Then tie the other end to any object. You'll know if anyone tries to take your treasure: pulling the string trips the trap. Use monofilament fishing line instead of string for an invisible alarm.

- You can also leave an audio reminder for someone special by arming their cell phone or car keys. They'll really get the message!

Personalized Talking Doll

The recording module from the Talking Booby Trap can be used in lots of other ways, like making this talking plush doll. Just record your own personalized sound or message on the circuit and stuff it into a doll. You can use an existing doll, or do what I did: create a custom doll from scratch.

For his Knetz Comics website, my brother Jack Knetzger created a talking corn kernel character named Niblet. I thought I'd surprise him by bringing his character to life in a talking doll.

I made the Niblet character using a simple two-piece "pillow doll" design. I made a pattern of the curved corn kernel shape and cut out two identical pieces of stretchy yellow cloth. Next, I pinned the cloth pieces together face-to-face and sewed them together around the edge, leaving a small opening. Then I turned the shape inside out and used fabric pens to draw the eyes and mouth. I stuffed the doll with polyester stuffing and placed the circuit inside.

Instead of using a bottle cap for the resonant chamber, I used a plastic cup the same diameter as the speaker so you could still hear the sound inside the doll's stuffing.

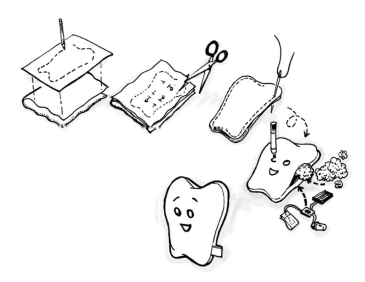

One fun detail: I made a cloth tag for the doll from the cartoon's logo by ink-jet printing the design on iron-on transfer paper and then ironing it onto a strip of cloth. I folded the cloth over and stitched the Play button inside the tag. (After recording the sound I snipped one of the leads to the Record button so that I wouldn't inadvertently press it again and erase it!)

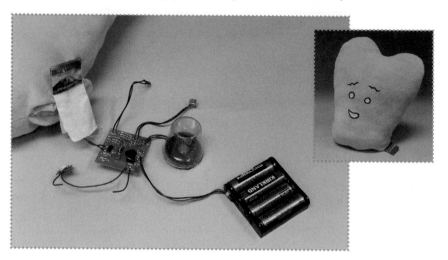

Then I sewed the tag into the seam of the Niblet doll and closed it up with some hand stitches.

When you squeeze the tag, the Niblet doll says his famous tag line "... but Lou, I'm just a kernel of corn!"

Scan this QR code to see a demo video of Niblet the Talking Doll in action!

You can create your own custom talking pillow doll by following my steps. Blow up a photo of a friend's face, ink-jet print it onto some T-shirt transfer material, and iron it onto a piece of cloth. Cut it out and sew it into a pillow, and stuff with stuffing and the module. Record a funny saying or sound effect and you can literally put words in your friend's mouth! Or make talking animal dolls or figures from your favorite comics, movies, or TV shows.

Breadboard 'Bots

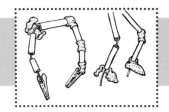

Many an electronic project starts on a protoboard, and for good reason. You can quickly wire up a circuit just by connecting resistors, integrated circuits (ICs), and other components using the push-in connector rails—no soldering needed! You can also easily disconnect and reconnect the wiring to revise your circuit as you go. For a little more fun, I added some Sugru® and some small plastic tubing to create these cute little Breadboard 'Bots.

This makes poseable electronic robot components. The Sugru acts as a flexible joint and holds the parts together. Thread some stripped 22-gauge copper bell wire through short pieces of ⅛″ plastic tubes and solder electronic parts to the wires. Add blobs of Sugru to make elbow, knee, and shoulder joints. Leave the wire sticking out the bottom of the feet for use with

protoboards. The alligator clip hands are wired together to grab and connect the 'Bot to another wire, component, or 'Bot.

You can make Breadboard 'Bots out of almost any component, like resistors, capacitors, and batteries, and it's especially fun when the component is an input or output device like a LED or mini speaker.

I thought it would be fun to make a 'Bot version of the photocell in Forrest M. Mims, III's classic 555 timer oscillator circuit. This 'Bot's photocell head is connected to a protoboard oscillator circuit through both his feet. Pose this Photo 'Bot facing up at you. When you shine a light on his face, the 555 timer circuit plays a squealing audio tone; the brighter the light the higher the pitch. Sweeeeeeeeeeet!

There's no limit to the size or look of your Breadboard 'Bots. Who says you can't have fun with functional circuits?

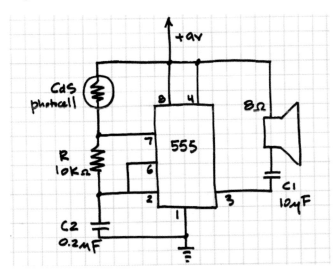

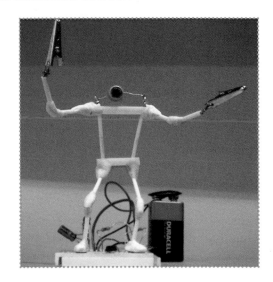

1980s Musical Toys

The 1980s brought big changes for electronic toys. Advances in integrated circuits meant electronics became inexpensive enough to be used in toys of all kinds, and especially in musical toys. Instead of plastic ukuleles or plinky pianos, these musical toys were something new, with new sounds, new methods of playing them, and new toyetic forms.

Playskool's cheap and cheerful entry into musical preschool toys was *Major Morgan*. He had a keypad with overlays that showed how to play a tune using color-coded grids. Just swap out his overlay for a new song. The rigid plastic keypad offered no tactile feedback (oww!) and the sound circuit produced just a single *BEEP* note at a time, but no matter: kids had fun with this musical soldier.

Mattel was aiming for an older crowd with their Bee Gees Rhythm Machine. This cute keyboard featured two disco-licious features: a pitch-bending wheel (*wow-weEEEee!*) and a toy version of a beat box with three different drum-and-bass loops. You could choose disco, Latin, or pop beats and adjust the tempo. The synthesized rhythms featured a busy bass line with a synth drum, and you could play along on the mini piano keys. This instantly

kitschy music toy was used by the group Kraftwerk on their hit song "Pocket Calculator" in 1981.

Mattel's *Star Maker Guitar* promised "hot" sounds but with only one string, musical choices were limited. You'd pluck the fat string and press it against the molded plastic frets to change pitch. The best part was a built-in "fuzz tone" effect for a fat, distortion-soaked sound. Unlike its real-world counterpart with magnetic pickups, this toy guitar had an *optical* pickup that "saw" the motion of the vibrating string. The speaker was mounted directly underneath the string producing endless feedback and long sustain. Young would-be Eddie Van Halens, shred on!

How about a drum set—without any drums? Nasta's *Hit Stix* were a big hit, that didn't actually hit anything. Kids could play "air drums" with a big sound. Inside the tip of each drumstick was an inertial switch that triggered the snare drum sound circuit when the stick was hit or shaken. It also triggered a trend in other "air instruments."

One of the weirder toys was Hasbro's *Body Rap*. Strap on an array of little switches and then slap your own thighs, ankles, wrists, and head to beat out a rhythm of sampled drums, cymbals, and even the spoken words *body* and *rap*. The 80s hairdo was not included.

内容：ドレミマジック（9V乾電池使

Perhaps the most iconic 1980s music toy was the *Magical Musical Thing* from Mattel. The TV commercial featured a kid playing melodies with one finger on the toy's keypad strip—and finishing by playing it with his head! The version sold in Japan showed a kid playing it with his butt—"*POOU!*" The toy's circuit was designed around the cheapest and most basic building block of digital electronics: the 4049 CMOS hex inverter chip.

Scan this QR code to see and hear these music toys in action!

Usually used for decoders and multiplexers, the lowly 4049 was reimagined by Mattel's thrifty engineers to create musical tones. Three of the IC's six logic gates were linked head-to-tail with a resistor and capacitor to create a simple, self-oscillating on-off, on-off square wave generator. The output of this oscillator was hooked up to the remaining three invertors in parallel. Their combined outputs were just enough to directly drive a speaker— no audio amplifier needed!

A network of resistors created the various musical tones. This was done using a cleverly designed membrane switch pad. The top and bottom layers were made from a single folded piece of mylar, printed with conductive silver traces connecting strips of resistive paint. The middle layer was a die-cut insulated spacer with holes positioned to make touch points, each labeled for a different color-coded musical note. Touching the membrane pressed together two conductive strips, which completed the circuit through a path of resistors, producing a single musical tone. The shorter the path, the less electrical resistance, the faster the circuit oscillates, the higher the pitch! *Beep—Boop!* Follow the color-coded notes to play a song or slide it over your body for a flourish of notes.

Musical Thing

Now you can make your own custom mini-version of this classic 80s toy. Wire up the circuit, draw a membrane keyboard, and make a housing (or put it in some repurposed container). This DIY redo has a new added feature: a circuit-bending touch point.

1. Wire the Circuit!

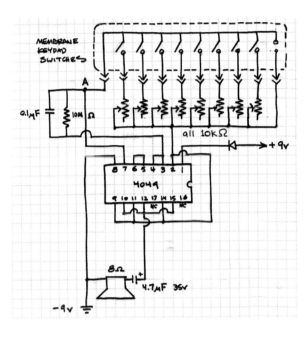

Materials

4049 hex inverter IC (plus a DIP socket, since the CMOS part is static sensitive)

0.1 uF capacitor

10M ohm resistor

Power diode

4.7 uF 12V capacitor

8 10K ohm mini trimmer pots

9V battery clip with leads

10 mini clips

Proto or perfboard

Wire

(Not shown: small 8 ohm speaker, solder, mylar, double-back tape, 3″ ribbon cable [10 conductor] membrane key pad templates—clip-and-use patterns in the appendix)

For housing:

Wood or foam for pattern

0.060″ styrene plastic sheet for vacuum forming

Tools

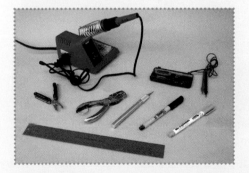

Soldering iron
Continuity tester
Craft knife
Straightedge ruler

Fine-point permanent ink pen
Conductive ink pen
Hand paper punch
Wire stripper

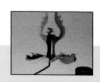

This reproduction circuit is quite simple with a minimum of components. You can easily solder it up with point-to-point wiring on a perf board. Layout isn't critical, but I placed the 8 trimmer pots in one neat row. After the board is done, cut it down to minimize its size.

Instead of the fixed printed resistors, this version has a trimmer pot for each of the 8 notes so you can tune them individually. Strip the ends of the ribbon cable. Solder the end of each of 8 wires to one of the trimmers. Solder a ninth wire to the circuit board for connection A. Solder a mini clip to the other end of each of the 10 wires. The tenth wire connects directly to pin 2—that's for the "circuit bending" pad. It doesn't need a trimmer; your fingertip provides the resistance value.

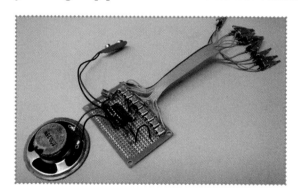

Solder the rest of the components following the wiring diagram.

2. Test and Tune!

Test the circuit: add the battery, and touch the clip at point A to each of the other clips.

You should hear a *BEEP* tone for each clip. Twist the trimmer pots to adjust the musical tones' pitch. Compare pitches from a piano or keyboard, or use a guitar tuner app to tune the 8 notes to a diatonic (do-re-mi) scale, like the white keys on a piano: C, D, E, F, G, A, B, C.

3. Make the Keypad!

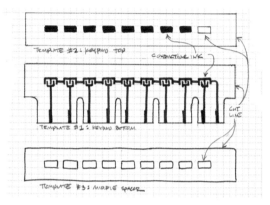

The membrane keypad has 3 layers, like a sandwich; you'll draw conductive traces on the upper and lower layers. The middle "meat" layer is just a spacer with some holes cut into it.

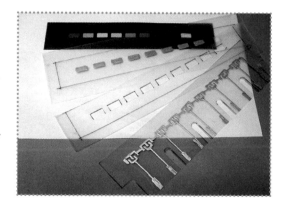

Find the clip-and-use templates in the appendix and cut them out. Tape a piece of mylar over the first template. If you're using frosted mylar, be sure to lay it frosted side down. Trace the cut outline with a thin permanent marker. Use the conductive ink pen to trace the circuit layout. Be sure to shake the pen well between strokes—some pens have a ball inside that works as an agitator. You'll have to very gently squeeze the barrel *and* press the valve tip down at the same time as you pull the pen across the mylar to make a uniform generous line. Join line segments while still wet for best conductivity.

Don't puddle the paint on too thick—it'll crack instead of flex when the mylar is curved. When the paint is dry, test each of the traces with a continuity tester and touch up with the conductive ink as needed.

Use the second template to make the other half of the keypad. Trace the cut-out

pattern with a permanent marker and color in the switch pads with the conductive ink.

The third template needs no conductive ink: just trace the outline and cut out the holes for the middle spacer layer.

To assemble the keypad, place the first layer with the conductive ink–side up and carefully align holes of the middle spacer layer over the first layer's switch pads. Add the topmost layer, with the conductive ink–side down, aligning the pads over the holes.

Tape the layers together temporarily and hook up the clips.

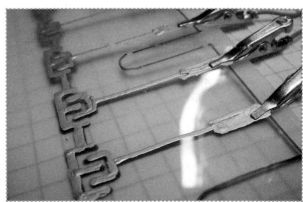

➜Closer Look: Conductive Pen Reviews◀

I found several different kinds of conductive ink pens that will work well with *The Electronic Connection*. Look for these in electronic stores or online. Here they are, starting with the best.

CIRCUIT SCRIBE BY ELECTRONINKS

These roller ball pens lay down a very controlled and fine line of silver-based ink. The ink dries really fast and conducts well. The pens are sold with kits that also feature magnetic electronic modules, like LEDs and switches, so even though they can cost more, you'll get more electronic parts to play with.

Tip: Store these pens vertically with the tips *down* for best ink flow.

SILVER CONDUCTIVE PEN BY MG CHEMICALS AND CIRCUITWRITER BY CAIG

These paint pen–type markers leave a thicker, painted line. They are a little harder to use and tricky to keep clog-free, but they give great results on any surface. They can be a little messy (you can clean up with nail polish remover), and take more time to dry, but the end result works really well on almost any surface. They are expensive (made with real silver!).

Tip: You must shake these pens really well before and during use. If they clog up, you can remove the tip for cleaning, but note: the caps twist off *backward* with a left-handed thread!

NICKEL CONDUCTIVE PEN BY MG CHEMICAL

A lower cost-version of the conductive ink pens that use nickel instead of silver. They require a similar application: squeeze to draw a thick paint-like line. It is not quite as conductive as the silver, but costs a lot less!

Tip: Remember to shake first—there's a small ball inside (like in a can of spray paint) to help mix the nickel paint and solvent.

ELECTRIC PAINT FROM BARE CONDUCTIVE

This product uses carbon in a black, water-based, paste-like paint. It is not as conductive as the silver or nickel, but it's not smelly and cleans up easily with water. It comes in a small squeeze tube so if you can't manage the tiny squeeze tip, try using a fine brush to draw lines.

GRAPHITE-FILLED CONDUCTIVE WIRE GLUE BY RADIO SHACK

This very thick substance is meant to be used as an adhesive that also conducts electricity, and is similar to the Electric Paint. It comes in a small tube with a tip applicator, but you might try a brush for better control.

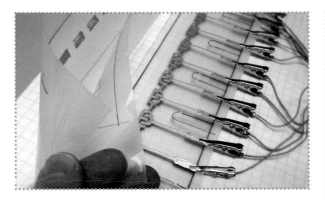

Press the switch pads to test your circuit. You should hear a beep each time you press and hold a switch pad. Again, touch up any traces or switch pads with a little extra conductive paint, if needed.

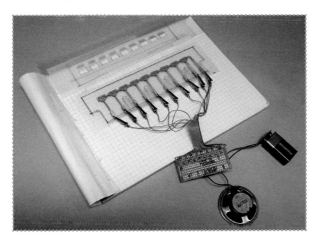

Trim the layers to size with a craft knife. Use the paper punch to make radiused inside corners as strain relief.

Use double-sided tape to fasten membrane switch layers together (don't put tape over any traces).

I added a colorful label on top with numbered touch points.

4. Play!

Reconnect the clips and play away! Adjust the trimmer pots to sweeten your tuning. Tune the eight notes to a do-re-mi scale as before for a useful set of notes. Or you can tune them to anything you want—including the first eight notes of any song for easy auto-play. Just swipe your finger across the keypad to play! Lick your fingertip and touch the last keypad position for some fun circuit bending sounds, from a low growl to a high squeal, and everything in between.

You can play with the Musical Thing circuit as-is, or put it in a project box or housing.

Making a Custom Housing

Here's how I made a custom molded housing for my Musical Thing. I wanted a miniature version of the original toy, so

I vacuum-formed some plastic shells to enclose the circuit.

I made a symmetrical wooden pattern and used it to vacuum-form two identical parts. I used one for the top housing and flipped a second one over for the bottom.

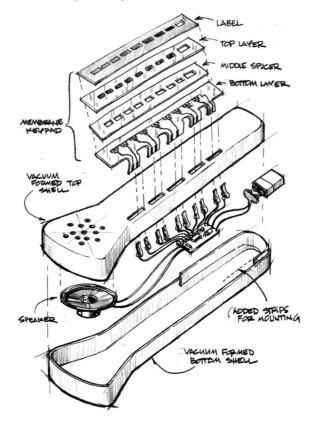

LABEL

TOP LAYER

MIDDLE SPACER

BOTTOM LAYER

MEMBRANE KEYPAD

VACUUM FORMED TOP SHELL

SPEAKER

(ADDED STRIPS FOR MOUNTING

VACUUM FORMED BOTTOM SHELL

Maker Tip

If you want to try this, see Chapter 4 for step-by-step instructions on how to make your own Kitchen Floor Vacuum Former. The appendix also has more examples of making custom molded housings using vacuum forming.

I marked the desired thickness with a surface gauge and trimmed off the excess.

I milled some slots in the top housing to thread the keypad connections through.

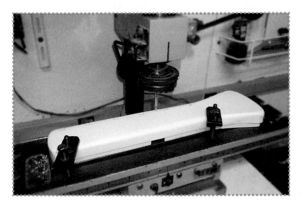

I drilled some sound holes for the speaker and then mounted the speaker, circuit board, and battery inside. I reinforced the rim of the top housing with some strips of plastic to make a lip all around the edge for the bottom housing to grip. I use methyl ethyl ketone (MEK) to quickly fasten the styrene parts together. See the appendix for more tips.

I painted the bottom housing a bright magenta. I taped the keypad in place using double-sided tape and attached the clips before closing the top and bottom shells. Looks and plays great!

The Electronic Connection

You can use the same sound circuit as in the Musical Thing project to make a different toy that plays conductive ink games. That's what Mattel did when they made *The Electronic Connection* in 1980.

Decades before Drawdio, Mattel's *The Electronic Connection* used an ordinary pencil as part of an electronic circuit to create a range of audio tones. The carbon in the pencil marks made on a piece of paper (and in the *pencil itself*) acted as a feedback resistor.

(Fun fact: the term *lead pencil* really is a misnomer; the stuff inside pencils is actually graphite, a form of carbon.)

The longer and skinnier the pencil lines, the higher the electrical resistance and the lower the audio tone created by the oscillator. Short, fat lines had more conductive carbon, giving a lower resistance and making higher-pitched tones. Combined with clever circuits printed on cards with conductive inks, *The Electronic Connection* let kids play electronic versions of pencil-and-paper games, including making music, spelling games, math quizzes, mazes, and more. I invented it in 1979—and even got a patent for it!

Build It!

To make your own version of *The Electronic Connection*, build the same electronic circuit as the Music Thing project with the same parts but eliminate all the trimmer pots and membrane keypad switches. Instead, just solder a wire with an alligator clip to the connection marked A and another wire with an alligator clip to the connection from pin 2 on the 4049 hex inverter chip.

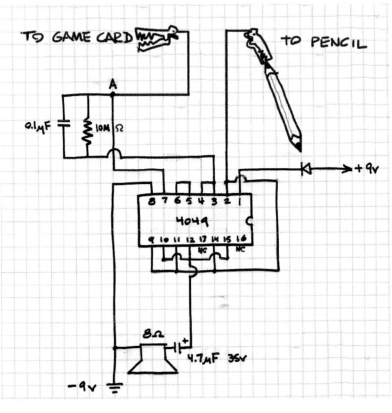

Maker Tip

You'll need a soft pencil to make and play with these conductive ink games. An ordinary number 2 pencil is too hard and won't leave enough carbon on the paper to complete the circuit. Look for drawing pencils marked 2B or 4B or even 6B at an art supply store. Staedtler Mars Lumograph 6B pencils work great!

Try It Out!

Sharpen *both* ends of a 2B (not 2H!) or softer pencil and attach one of the alligator clips to the pencil. Draw a thick, fat patch of pencil-marking near the edge of a piece of paper. Connect the free alligator clip directly on that mark. When you touch the pencil tip to the pencil mark you complete the circuit and hear an audio tone: *BEEEEEP!* Make crazy sound effects from a low growl to a high squeal as you draw with the pencil and add more and more conductive carbon lines to the circuit: *zzzzzeeeEEEEEE!* Have fun making sounds and drawing paths and pictures. You can also use your fingers to touch and complete the circuit. Hold the pencil tip in one hand and touch the pencil marks on the paper. Just making the sounds with the circuit is fun by itself—but there's more.

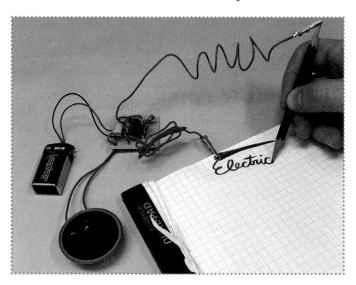

More Fun with Game Cards!

The next step is to create game cards using conductive ink pens to draw circuits. Find and cut out the clip-and-use cards in the appendix. For a longer-lasting card, laminate the paper onto thin

cardboard (like a file folder) with some glue or double-sided tape, then cut out.

Color in the blue lines with conductive ink. Let the ink dry completely, then color in the red areas with pencil. Press down hard for a dark pencil line. Use a 2B or softer pencil and press firmly to fill in the resistance values, as shown in red, with a nice thick layer of pencil. Hook up the alligator clips and use the pencil as a stylus to play the games.

Try these fun electronic circuit game cards.

DEXTERITY SKI RUN

Here's how to play this easy skill game: start your pencil probe on the center line at the Start. You'll hear a low tone—that's good. Now trace the path of the center line toward the Finish. Don't press or draw; just touch lightly with the pencil tip to keep the tone playing. Be careful! If you stray or lift the pencil, the tone stops: that's a "fall." Make a check mark in the Fall box, go back to Start, and go again. If you really stray, you'll hear a high-pitched tone: that's a double fall. Make two marks in the score box, go back to Start, and try again. What's the fewest number of falls you can make to get to the Finish? Can you do a perfect run? Think it's too easy? Hold the pencil in your *other* hand!

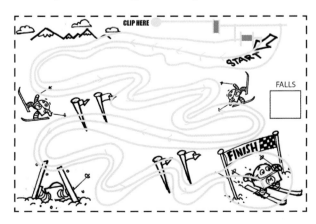

3-IN-1 SOUND MAZE

This game is a maze that magically changes sound to direct you to one of three different goals. First, pencil in one of three goal circles. Use the pencil to completely fill in one of the three circles to make an electronic connection. Go to Start and lightly touch the pencil tip as you trace a path through the maze. Listen to the buzzing tone. As long as the tone you hear keeps getting higher, you're going the right way. If the tone you hear starts to fall, that means you're going the wrong way. Retrace your path and keep going, following the rising tones until you get to Finish. Then, erase the goal circle and color in a different one: the maze changes sounds as you trace to the new goal.

2-PLAYER BASEBALL

Here's a baseball game for two players with hits, outs, and innings.

At the start of the inning one player secretly chooses five of the "hit" baseballs and colors in their five red squares with pencil. Each filled-in ball will make a *buzz* sound when touched with the pencil. Then fold the paper over to hide the colored-in choices. The second player "bats" by choosing a ball and touching it with the probe. If there is no tone, it's a hit! Mark the base runners on the diamond lightly with the pencil and bat again by choosing a different ball. But if you hear a buzzing sound, that's an out! Update the scoreboard and keep playing. Keep tallying bases, runs, and outs just like in real baseball. After three outs, players switch sides: now the first player will bat after the second player erases and colors in five red squares. Will he fake out his opponent and keep some of the previously colored hits as outs again? That's up to you . . . play ball! You can play nine whole innings of conductive ink fun.

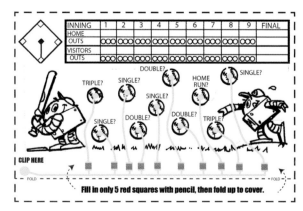

MAKEY BOT MUSIC

Three Makey bots are playing some music: a low-pitched bass, a sliding trombone, and a high-wailin' sax!

Trace the blue lines with conductive ink, then color over the red areas with pencil. Hook up the alligator clip to the card. Touch the pencil tip to the spots on the various instruments. The bass plays low notes. The sax plays high notes. Slide the tip along the trombone to make a sliding trombone sound! Slide your way across the trombone to play a melody.

CROSS-GRID STRATEGY GAME

This is a strategy game for two players. The first player to complete a continuous line that connects all the way across the grid and sounds the buzzer is the winner!

On each player's turn, he or she draws a line connecting two of their shapes. (Press *firmly* to draw solid pencil marks along the dotted lines.) The "circle" player draws on the dotted lines connecting any two adjacent circles. The other player, playing "squares," draws on the dotted lines connecting any two adjacent squares. No diagonal lines are allowed. Players cannot cross their opponent's lines. Keep taking turns coloring in lines until one player has drawn a continuous line connecting all the way across the grid.

To claim a win, touch the pencil tip to your shape's "Win Test" spot. If you hear a buzz or growl tone (no matter how low-pitched) you win!

Carefully erase all the lines to play again.

Chapter 3

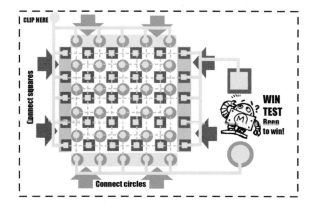

Exploring Science with Toys

*E*ven the most basic playthings like spinning tops, floating soap bubbles, and bouncing balls illustrate the principles of science in action—if you look carefully. You can observe and learn about physics, chemistry, electricity, hydrodynamics, pneumatics and hydraulics, magnetism, basic machines, optics, and on and on.

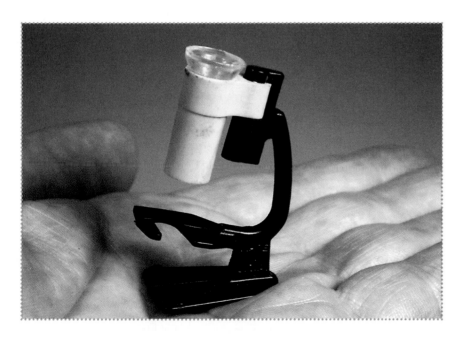

FREE! INSIDE!
Science in Your Cereal Bowl!

Cereal manufacturers learned early on that an inexpensive giveaway inside a box of cereal could inspire a purchase and create brand loyalty with moms and kids. From the 1950s on, the cereal aisle in the grocery store became a mini toyshop with an endless array of plastic baubles, punch-out character masks, and collectible trinkets. Amid the secret decoder rings and cowboy sheriff badges were some cleverly designed toys that ingeniously used scientific principles to amaze and entertain.

Hey Kids! Cereal Science Toys Build Inquisitive Minds 6 Ways!

#1. Frosted Flakes Diving Tony

This cereal science toy was a witty 1981 remake of a classic scientific toy, the Cartesian Diver. This miniature version of Tony the Tiger mysteriously obeys your commands as he dives and rises inside of a water-filled soda bottle. Grrreat!—but how did it work?

The scientific secret used the incompressibility of water along with the ideal gas law:

$$PV = nRT$$

In short: The volume of a gas is inversely proportional to the pressure on it. The plastic Tony was molded to be neutrally buoyant and to float near the top of the bottle. When you squeeze the bottle, the pressure on the water compresses the air bubble inside Tony. The reduced bubble displaces less water, making Tony less buoyant, and he sinks to the bottom. When you release the pressure, the bubble expands and displaces more water. Tony becomes buoyant again and rises back up. Because you can't see the bottle being squeezed, the up and down diving action seems magical! You could even subtly control your squeeze to make Tony pause and float at any depth. The side of the cereal box had a matching graphic of a deep-sea dive game complete with a wrecked ship, sunken treasure, and a menacing shark—also grrreat!

But one mystery remains: why is the Cartesian Diver named for the wrong Frenchman—René Descartes, instead of Blaise Pascal, the 17th-century father of hydraulics? Maybe Tony is trying to tell us: "I sink, therefore I am!"

Today toy collectors covet Diving Tony and he's hard to find; but you can build your version at home!

#2. B'sun Whistle Phone Hacker

How did a simple cereal-prize whistle empower an early phone hacker and inspire the founders of Apple Computer? In 1971 Quaker Oats packed FREE "Bo'sun Whistles" inside boxes of Cap'n Crunch cereal. Molded in bright colors and embossed with the Cap'n and Seadog, this giveaway produced a piercing two-toned blast which, like a real bosun's whistle, could be heard over the sounds of the sea or in bad weather.

Coincidentally, the whistle's perfectly pitched 2600Hz tone could *also* be heard by AT&T's analog telephone trunk-line switching circuits. Tricking the billing circuits resulted in FREE (well, stolen) long-distance phone calls for anyone who knew the illegal secret. Phone phreaker and computer programmer John Draper, nicknamed "Captain Crunch" for infamously demonstrating this slick whistle trick, went on to develop electronic tone-generating circuits to do the same thing. Draper's technical skills impressed a young Steve Wozniak, who hired him to create circuits for Apple Computer. Unfortunately, publicity from a magazine article "blew the whistle" on Draper and he was convicted on toll fraud charges. He wrote the first word processing software for the Apple][while in "the brig." Today, Captain John Draper is FREE, and modern digital phone circuits are unaffected by these nautical noisemakers.

#3. Balloon-Powered Rocket Cars

A less controversial air-powered cereal toy was the balloon-powered car. There were many versions of the toy, but they all obeyed Newton's third law of motion, the mutual forces of action and reaction between two bodies. The force created by a jet of air escaping from a rubber balloon was powerful enough to propel a toy car in the opposite direction. As early as 1950, Kellogg's offered a Jet-Drive Whistle Loco available by mail for 25 cents and a box top from Kellogg's Corn Flakes. According to the promotional literature at the time, the four-inch-long injection molded plastic steam locomotive was "Accurately scaled—even the rivets show)." The corny sales pitch to grocers touted "All steamed up—and ready to go! Get aboard— here's your ticket to biggest sales yet! Is there a kid in America who *wouldn't* want one?" Only the mailman knew for sure.

This science-based gimmick was an irresistible force that kept coming back. In 1961, Nabisco's Rice and Wheat Honeys cereals came with updated Racing Robot and Speeding Spaceman variations. By the 1970s, Quaker Oats included their own version: Balloon Racers, FREE inside boxes of cereal. The more compact 2½″ design kept plastic costs *down*, and, thanks to Newton's second law of motion, Force = (mass) × (acceleration), kept car speeds *up*. With less mass, these smaller cars featuring the Cap'n, Jean LaFoote, and Smedley the Peanut Butter Crunch elephant went even faster on the same balloon power!

#4. Mighty Mini-Magnifiers

The science of optics has also proved to be a popular platform for premiums. Periscopes, spyglasses, and lenticular lens "wiggle pictures" have all been offered as fun freebies. You almost need a real magnifying glass just to see the real working mini-microscope, free in Wheaties! More of a miniature magnifying glass, the single plastic lens in an adjustable

eye-piece tube afforded very limited magnification and plenty of optical distortion—but it really worked!

#5. Give-Away Gyroscope

The science behind even the simplest toy top or yo-yo could make your head spin. The kinetic energy of a flywheel is described by the formula:

$$E_k = \tfrac{1}{2} I\omega^2$$

where E is the amount of kinetic energy, ω is the angular velocity (how fast it's spinning), and I is the moment of inertia (resistance to change in spin). To find the moment of inertia for a disc shape with a large center hole, just use $I = \frac{1}{2} m(r_1^2 + r_2^2)$. Got that?

Fortunately no math skills were required to have fun with Quisp cereal's Gyro-Cycle or Flywheel Car premiums. One tug of the rack-and-pinion geared pull strip and the mini flywheel instantly revved up to high speed (plenty of ω!). The clever design also included a metal disc in the flywheel (much more mass than a plastic one) for increased I. All that ω and I resulted in enough kinetic energy to send Quisp, the spin-powered spaceman, quickly zipping across the floor.

#6. Bubble-Powered Sub

One classic cereal toy has stayed crunchy in the milk of time for over 50 years: the baking powder–powered diving sub. In 1955, boxes of Kellogg's Rice Krispies promised "FREE INSIDE! An actual working ATOM SUB!"

In reality, a simple kitchen chemical reaction provided the fuel for millions of these miniature marvels. A pinch of baking powder from Mom's pantry contained both an alkaline, sodium bicarbonate ($NaHCO_3$), and an acid, cream of tartar (potassium bitartrate, $KC_4H_5O_6$). When combined in water, the two react:

$$NaHCO3 + KHC_4H_4O_6 \rightarrow KNaC_4H_4O_6 + H_2O + CO_2$$

The reaction produces a salt, water, and carbon dioxide gas: CO_2. In recipes, the CO_2

makes quick breads and muffins rise. In this case, the CO_2 makes toy submarines rise. The buoyant force of the bubbles formed is enough to lift the submerged sub. When it surfaces, the sub tips to one side, the bubbles are released, the sub sinks, and the cycle repeats.

Over the years, there were many versions of this buoyant bubbling toy: submarines, diving frogmen in assorted sizes, killer whales, sharks, and mechanical monsters. Although the most famous was the 1955 submarine design, created by the brothers Benjamin and Henry Hirsch, a patent search finds an even earlier design dating from 1920. Who knows how many other versions are floating around?

Hey, Kids! Now you can make your OWN version of the famous diving sub—no box tops needed! Just ask mom for a potato to make your own working version of the diving sub.

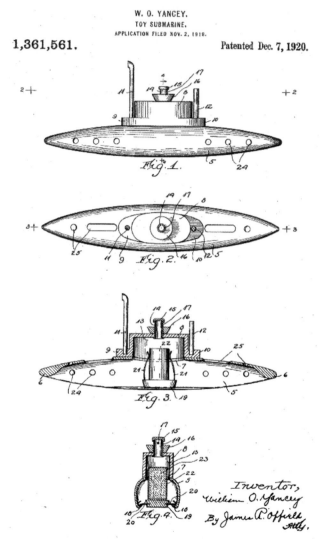

Cartesian Diver

I used a short piece of clear PETG tubing with a matching vinyl cap. I also cut a piece of solid styrene rod that fit snugly inside the tube. File or sand a small flat area along the length of the rod. This provides a channel for the water to compress the bubble of air up inside the cap. Cap the tube and insert the rod. Test the diver in a sink of water for neutral buoyancy. Place the diver in the water with the cap facing up so that a bubble of air is trapped up in the cap. Slide the rod in or out to adjust the size of the air bubble inside until the diver just *barely* floats to the top.

Fill a one-liter plastic bottle to the top with water, insert your diver, and cap tightly. When you squeeze the bottle, the water pressure compresses the air bubble, which then displaces less water, and the diver sinks. When you release, the bubble expands, and the diver rises. With a little practice you can make your Cartesian Diver obediently rise or dive on your command.

Materials

½" diameter PETG tubing, 2¼" long
½" vinyl cap; fits snugly on the tubing
½" plastic rod, 2" long
File or sandpaper
Hobby knife
Saw

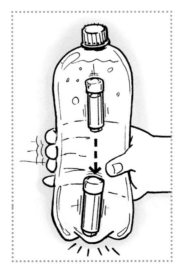

vinyl cap

1/2" D PETG tube

styrene rod w/ ground flat

Scan the code to see the Cartesian Diver do its thing!

If you like, add a scuba diver or octopus graphic from the clip-and-use pages in the appendix.

Diving Spudmarine

First, cut a potato into a ¾″ diameter cylinder, about 3 inches long. Or, just use a small fingerling potato about that size. Use a piece of ¼″ brass tubing as a plug cutter. Press the tube all the way into the potato to make three through-holes along its length. This will reduce the mass of the potato for better diving.

Enlarge the bottom of the center hole to make a flared conical opening. This will create an air chamber for the bubbles.

Next, make a periscope. Cut a thin piece of wood about ½″ by 1″ and drill a ¼″ hole in the center. Cut a 1″-long piece of ¼″ wood dowel and insert it in the hole.

Place the periscope into the middle hole in the potato.

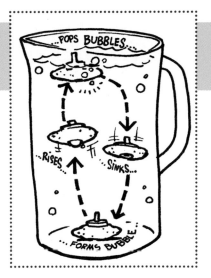

Time to test: place the Spudmarine into a tall pitcher or vase of water. If it floats, cut off some of the periscope and try again. If it sinks FAST, trim off some of the potato and try again. If it sinks very s..l..o..w..l..y, it's ready! The sub should be just ever-so-slightly heavier than neutral buoyancy for best diving action.

Remove the sub and shake it dry. Pack some baking powder (not baking *soda*!) into the bottom of the center hole. Use some more wood dowel to tamp it in tightly.

Gently lower the sub back into the water and let it sink, then watch it bubble . . . rise . . . breach the surface and burp its bubbles . . . and then sink again! How many times will your sub rise and sink before you have to reload the powder?

Victorian Toy and Flatland Rockets

Materials and Tools

- Paper fish clip-and-use from the appendix
- Lid from margarine or yogurt container
- Shallow pan or cookie sheet with a rim all around
- Water
- Olive oil
- Dish detergent
- Toothpick, drinking straw, or eyedropper
- Scissors
- ¼" paper punch

Here's a Victorian-era plaything: a fish that swims by the power of surface tension—and a modern version of the toy that uses a special floating plastic.

1. Try a Quick Paper Version!

Find the fish images in the appendix and cut out on the dotted line. Fill a small pan or cookie sheet with clean water and carefully float a paper fish in the middle of the pan. With an eyedropper, drinking straw, or toothpick, place a single drop of olive oil in the fish's cut-out circle. The oil quickly spreads out the slit and across the water. The fish "swims" in the opposite direction, like an exhaust-spewing rocket subject to Newton's third law of motion. Sadly, a soggy paper fish is only good for a single use.

2. Plastics to the Rescue!

Now try this new, more durable version: find a flexible lid from a margarine or yogurt container. Look for the recycling symbol 2 or 4 for low- or high-density polyethylene (PE). Fun fact: PE is the one of the few non-foamed plastics that *floats!*

Use a paper punch to make a small circular hole in the lid, then cut out the "rocket" shape from the clip-and-use version in the appendix.

Float the rocket in a pan of clean water. Dip the tip of a toothpick in detergent and momentarily touch it inside the rocket's round hole. As the detergent dissolves, it spreads down the slit and out along the surface of the water—the rocket shoots forward! Touch it again. *Zoom!*

Another force is also at work: the *Marangoni effect*, the difference in surface tensions created by the molecules of detergent as they make the water slipperier and "wetter." The surface tension is reduced behind the rocket, causing the water in front to contract, pulling the rocket forward.

These tensions, forces, and actions all exist at the single-molecule-thick surface of the water—similar to the two-dimensional world in Edwin Abbott's Victorian-era book, *Flatland: A Romance of Many Dimensions*! Your margarine-lid rocket only works in two-dimensional space. The waterproof plastic rocket lasts much longer than the soggy paper fish. After a few uses, though, you'll have to change the water for the effect to work again.

Rocco and Violet Marino launching some Flatland Rockets. Photo courtesy Robert Marino.

Scan this QR code to see the *Flatland* rocket in action.

Good Vibrations:
Groovy Mechanical Sound Players

Long before iPods, mp3s, or even electricity, people recorded and listened to music and speech by all-mechanical, analog means. Thomas Edison's first important invention after setting up shop in Menlo Park was the practical phonograph. His 1877 design featured a sharp stylus that pressed into a tinfoil cylinder. When he shouted into a horn as he turned a crank, the vibrations of his voice made a pattern of indentations in a corkscrew groove along the surface of the spinning cylinder. When the stylus retraced that up-and-down pattern, the vibrations reproduced the sound of his voice. Edison famously demonstrated the effect by reciting the nursery rhyme "Mary Had a Little Lamb" and then playing it back.

It's telling that he chose so frivolous a passage. Although the original intent of the invention was not for entertainment (Edison was working on a way to record telegraph messages), the first application for the phonograph was in a talking doll. Unfortunately the fragile, bisque-headed dolls proved too expensive, unreliable, and difficult for children to operate (you had to turn a crank smoothly and continuously to make the doll talk). The doll was a commercial flop and Edison moved on to other projects.

By the early 1900s, others had made improvements on the phonograph by replacing the foil with wax, incising rather than impressing the grooves, and switching from individually recorded cylinders to discs that could easily be duplicated by stamping. Each of the two incompatible formats had their fans. Some argued that the constant stylus speed across a cylinder together with the "hill-and-dale" modulation (grooves that wiggled up and down) of Edison cylinders reproduced sound better than disc format with sideways modulation, varying stylus speed, and tone arm tracking error. No matter—like the Betamax/VHS showdown, the technically inferior but more popular gramophone disc format prevailed. Then with the advent of radio and electronically amplified phonographs, wind-up mechanical sound players died out . . . until 50 years later.

In 1960, Mattel introduced a new talking doll that was everything Edison's doll wasn't. *Chatty Cathy* was sturdy, great sounding, affordable, and most importantly, easy for a child to operate. Just pull the string and Cathy said one of 11 different phrases, like "Tell me a story!" or "Please take me with you!" Mattel gave her a soft vinyl head with rooted hair and accessories like strollers and a wardrobe of themed outfits, all "sold separately" of course. Thanks to her nationwide TV campaign, *Chatty Cathy* was a big success.

What made it all work was the cleverly designed voice unit, which was patented by ex-Raytheon missile engineer Jack Ryan, Mattel's in-house toy wizard.

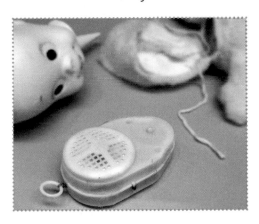

Instead of turning a crank, the child simply pulled out a string, which wound up a powerful metal spring. In the same motion the string (ingeniously threaded right through a hole in the tone arm) automatically lifted and pulled the tone arm back to the beginning of the record.

The miniature record had concentric annuli. Unlike a single continuous groove (like on an LP), the disc had multiple grooves that were interleaved and spiraled around each other. They were arranged so that the multiple lead-in grooves of the tracks were distributed around the rim, like numerals on a clock face. When the tone arm dropped on the spinning record it would land at random on any one of the tracks. "You never know what she will say next!"

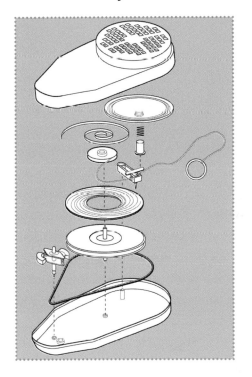

Unlike in the gramophone, the styrene speaker cone in this voice unit was fixed in one position. A tiny spring-loaded piston pressed the needle into the hill-and-dale modulated groove. It also acoustically coupled the speaker cone to the curved ridge on top of the moving T-shaped tone arm. The record was molded out of tough, slippery nylon to be both durable and smooth running. This design produced *loud*, clear sounds with great fidelity. The constant pressure of the piston kept the needle in the groove so the voice unit didn't rely on gravity like a gramophone—it would work upside down, or at any angle, perfect for a toy.

TV's *The Funny Company* fluttered her eyelashes and moved her lips.

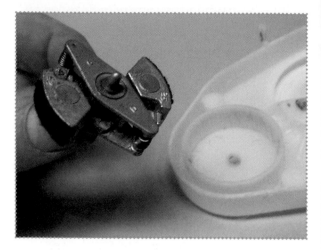

To keep the record spinning at a smooth and steady speed, it was connected via a rubber belt and pulley to a die-cast zinc centrifugal regulator. Like a spinning figure skater, the arms of the regulator would move in and out in response to any variation in speed. Too slow and the spring-loaded arms snapped inward, speeding it up. Too fast and the arms swung out, where their felted tips dragged against the housing, subtly and gently braking the speed.

Mattel continued to improve their voice unit by updating to a more elegant S-shaped negator spring, which provided constant force in a compact size. The powerful spring motor had extra torque, which was used to power additional mechanical gimmicks on various toys. *Mickey Mouse Chatter Chum* moved his head up and down as he talked. *Talking Shrinkin' Violette*, a doll based on ABC

The same basic mechanical sound player was used in dozens and dozens of other toys: Barbie dolls, talking books and games, puppets, and lots more. Because these miniature sound-makers reproduced recognizable voices and sounds, they were a natural for items that were based on well-known characters with famous catch phrases. Just pull the string to hear Robin Williams as Mork from Ork say, "Nanoo, Nanoo!," Herschel Bernardi as Charlie the Tuna say, "Hey, Stahkist, I gaht good taste!" or Mel Blanc as

Bugs Bunny say, "What's up, doc?" The list of character voice toys was endless: Casper the Friendly Ghost, Beany and Cecil, Doctor Doolittle, The Monkees, Herman Munster, Woody Woodpecker, Flip Wilson, Fred Flintstone, and many more.

The pull-my-string action was so beloved that it was used to trigger the *Toy Story Talking Woody* doll, even though the toy's sound player was entirely electronic.

The longest-lived pull-string product line was the preschool *See 'N Say*. A pointer attached to the record's shaft allowed a kid to select the particular sound they wanted to hear. Just point and pull to hear nursery rhymes, letters of the alphabet, numbers, or animal sounds. "The cow says mmmMMOOooooo!" Today, the iconic *Farmer Says See 'N Say* talking toy lives on in a smaller, cost-reduced version. The string still pulls out and the pointer still spins, but all the sounds are produced electronically.

Even after the 1970s when all-electronic talking toys were introduced, these mechanical players offered an inexpensive way to reproduce natural voices and sound effects. One toy from 1982, Mattel's *Teach & Learn Computer*, combined Victorian and Space Age technologies. A microprocessor was used to accurately drop the tone arm onto a spinning record, landing at the exact instant to play the desired single track out of 40 different lead-in grooves whirling by.

THE TLC TONE ARM (IN WHITE) IS RELEASED BY THE SMALL MOTOR (THE GRAY CUBE TO THE RIGHT) CONTROLLED BY THE ELECTRONICS. A TLC INTERCHANGEABLE RECORD WITH 40 DIFFERENT RECORDINGS, EACH TRACK WITH ITS OWN LEAD-IN GROOVE.

Mechanical sound players continue on in novelty applications. Scan the QR codes to see some videos of a Japanese all-cardboard record player toy and thumbnail-activated talking strips. Hill-and-dale modulation lives on!

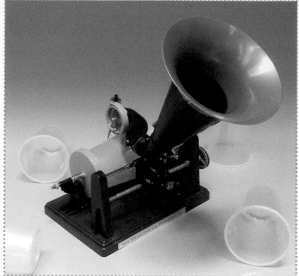

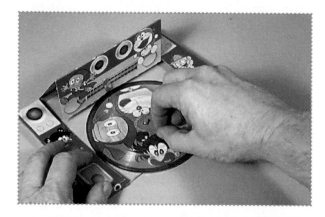

And coming full circle, you can currently buy plastic reproduction kits of all-mechanical phonographs. Gakken, the Japanese science kit maker, has cleverly updated Edison's phonograph design to use plastic drinking cups instead of tinfoil cylinders and it's powered by a small electric motor instead of a hand crank. There's also a Berliner Gramophone Kit disc cutter with a wind-up motor. See the Maker Shed online for the latest offerings.

Groovy, baby!

Voice-Powered Pigs

You never know where the next new toy or game idea may come from. Years ago, I was sitting in a lecture hall listening to a boring speaker drone on and on. ZZzzzzz…. That droning sound was the inspiration for a new game idea. I wondered: could I use the vibrations from the sound of my voice to move tokens across a track? I couldn't wait to get home and test it out using some pieces from an old electric football game on a thin, resonant surface—it worked!

So I made a four-player "pig race" prototype: each player yelled into a flexible tube that directed the sound to a stiff and lightweight styrene track. As the track vibrated, the pig tokens moved along on little bristles. Each player had his or her own track, which all ended in a center finish line. The players yelled "sooo-eeeey!" and their pigs skittered and jiggled along toward the finish. Yell the most to win the race!

Hog Holler was produced in 1990 and was featured on *The Tonight Show*. Johnny Carson and Ed McMahon hilariously hooted and *sooo-eeeey-ed* to demo the game on network TV—a long, long way from the boring lecture hall!

Scan the topmost QR code to see the commercial for *Hog Holler*.

Scan the lower QR code to see a clip of *Hog Holler* on *The Tonight Show*.

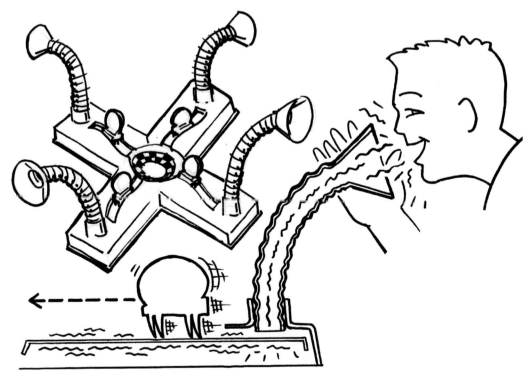

Cereal Box Sound Racers

Build your own version of this voice-powered race game. Hoot and holler through the tubes to send the tiny tokens skittering across the cereal box. First token to get to the edge and fall off wins!

1. Cut

Using the hobby knife, cut a thin disc from the cork. It should be about ⅛" thick.

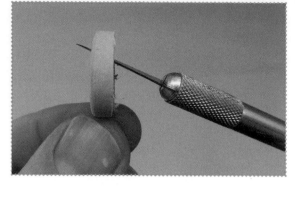

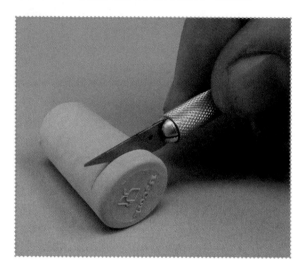

Use the tip of the knife to pierce through the cork disc to make three slots. Make the slits all at the same slight angle, about 15° degrees from perpendicular.

2. Assemble

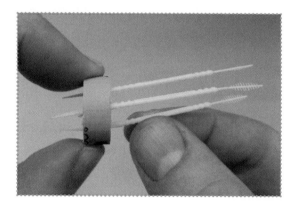

Insert the pointed end of a pick through the slit until it sticks out about ¼". Insert the other two picks in the remaining two slits. Adjust the picks to stick out the same length

Materials

Empty cereal box
Cardboard tubes from a roll of paper towels
Wine corks (the synthetic kind, not real cork)

Plastic toothpicks (use molded plastic dental pics; the kind with a thin, blade-shaped, cross-section)
Hobby knife with #11 pointed blade
Side cutters

to make three little angled legs to support the disc. Gently spread the two rear legs apart as shown. Snip the other ends of the picks, leaving just the legs. Adjust the legs as needed so the disc stands level.

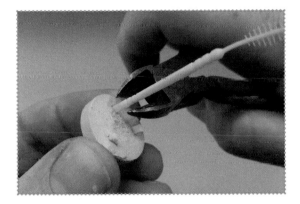

SIDE VIEW WITH ANGLED LEGS.

VIEW FROM BACK WITH SPREAD REAR LEGS.

Make a few more tokens from the corks and picks. Mark the tokens with colored marker.

3. Race!

Place the cereal box flat on the tabletop. Each player chooses a token and puts it together with the others on the middle of the box. Gently touch a paper cardboard tube to the top surface of the box but away from the tokens. On the signal all players make a long "WHOOOOOooooo" sound into the tube. Try different pitches high or low, loud or LOUDER to find the best resonant frequency to make the tokens skitter along. Keep hooting until one token gets to the edge of the box and falls off—the winner!

MAKER TIP

Adjust the legs for best performance. Try slightly different angles for the legs, or different amounts of rear leg spread. Tweaking individual legs can change the direction the token will go: in a circle, left, right, or straight ahead.

Mastering 3D Views!

With ever more 3D movies and virtual reality viewers like Oculus Rift, three-dimensional entertainment is hot! Here's a look back at a clever design for taking and viewing stereo photography *long* before digital cameras and smartphones.

Wouldn't it be fun if you could take your own family pictures in thrilling 3D? That's what William Gruber thought back in 1939. Although stereo photography had been around for years, creating your own 3D photos was more than a little complicated. It required

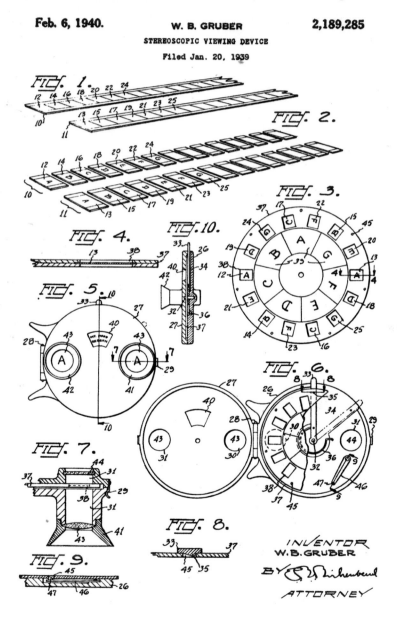

Feb. 6, 1940.

W. B. GRUBER

2,189,285

STEREOSCOPIC VIEWING DEVICE

Filed Jan. 20, 1939

INVENTOR
W. B. GRUBER

BY
ATTORNEY

special cameras and lots of technical know-how to take the two images that would "fuse" to create a deep 3D space. That was fine for hard-core hobbyists, but what about the average person? For them, Gruber elegantly combined several elements to create View-Master, a system for making and viewing 3D.

Gruber realized that the then-new Kodachrome 16mm film could be used for more than just movies. The long strip of film could be cut into tiny, individual frames—an economical way to make millions of vibrantly colored transparencies.

Gruber then laid out the frames as stereo pairs in a ring around the edge of a flat disc. The intraocular distance (the distance between your eyes) determined the disc's diameter at 3 inches. A disc that size (called a "reel") was just big enough to hold seven pairs of 16mm frames of film. Voilà!—the View-Master's haiku-like sequence of seven images per reel was born.

The first View-Master viewer was made of brittle phenolic plastic with a split hinge that opened to change reels. As you pulled down on the lever, the reel would swivel on a center pin, swinging the next image into view with a snappy sound. That "tug, swirl, clack—Wow!" is the well-known and beloved View-Master 3D viewing experience. (To this day, View-Master reels are still made with that same punched center hole, even though it hasn't been used in viewers in the last 60 years! How's *that* for backward compatibility?)

Forget Grandma's klunky black-and-white stereopticon cards. Now everyone could easily view colorful 3D pictures anywhere. View-Master went on to make and sell more than 1.5 billion reels with images from around the world, scenic wonders, coronations, animals, fairy tales, and more. Like the ad says, "the breath-taking beauty of View-Master pictures is a new and delightful experience."

A PARADE OF SELECTED VIEW-MASTER VIEWERS (L TO R): MODEL B 1940S; MODEL C WITH LIGHT ATTACHMENT, 1950S; MODEL D WITH ADJUSTABLE FOCUS, 1960S; MODEL M WITH PUSH BUTTON ADVANCE, 1980S; MODEL "VIRTUAL VIEWER" WITH LARGE LENSES; AND AT TOP, THE CLASSIC MODEL L, STILL MADE TODAY. AS OF THIS WRITING AN ALL-NEW VIEW-MASTER IS BEING DEVELOPED: MATTEL HAS REENVISIONED GOOGLE'S CARDBOARD SMARTPHONE HOLDER INTO WHAT PROMISES TO BE AN ENGAGING, INTERACTIVE, VIRTUAL-REALITY VIEWER.

But how do you take your own 3D View-Master pictures? The second part of Gruber's system was the 1950 View-Master Personal Stereo Camera. It used Kodachrome slide film for brilliant color and smooth, grain-free results. But where other stereo cameras took full-frame 35mm images that required complicated mounting (more about that later), the View-Master Personal Stereo Camera created the same 16mm-size images to fit the standard View-Master reel. That meant only half of the width of the 35mm filmstrip was used.

Gruber's aha moment came when he envisioned running the film through the camera *twice*: once as the film was pulled out of the canister, then again as the film was wound back into the canister. What made it possible was the *film miser*, an elegant solution with movable lenses.

At the end of the roll you'd simply twist a dial.

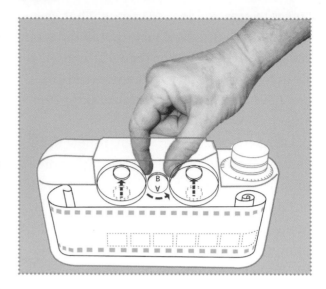

The swiveling twin lenses would be repositioned from the bottom half of the filmstrip to the top where they would make a second strip of stereo images.

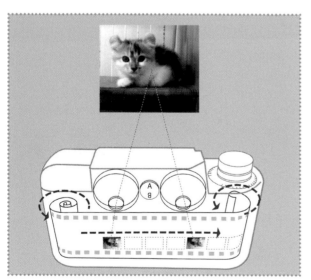

With the dial set to "A" the camera takes images along the bottom half of the film as it's wound *out* of the canister.

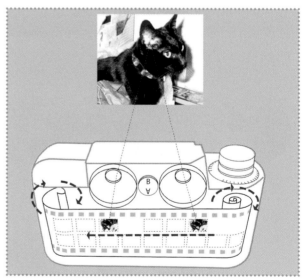

Using a 36-exposure roll of film, the View-Master Personal Stereo Camera yielded a whopping 69 stereo images. Wow!

With lots of other clever features, the View-Master Personal Stereo Camera was a breeze to use.

A bubble level appeared right in the viewfinder to help you hold the camera level. When taking pictures in "you-are-there" 3D, you don't want sea-sickening tilted horizons.

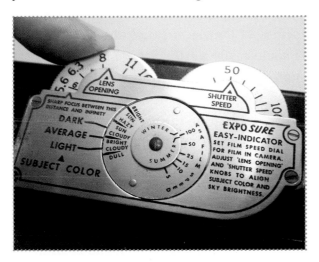

Built into the top of the camera was the Expo-Sure, a clever light-metering system. First, you set the speed of the film you were using (100 ASA was the top speed setting!) and select the season (winter or summer lighting). To take a picture, you set the f-stop and shutter speed control knobs to match up the subject color (marked dark, average, and light) to the sky brightness (bright/sun through cloudy/dull). When you lined up the marks, the f-stop and shutter speed were automatically set for a perfectly exposed picture. No light meter needed. It also indicated the depth of field right on the dial. Did you want more of the picture in deep focus? You would just use a smaller aperture. The Expo-Sure automatically adjusted for the new

shutter speed setting—it was an all-mechanical analog computer!

The camera's focus was fixed, but for close-up subjects, a snap-on lens attachment would add magnification and slightly change the angle of the two lenses—like crossing your eyes to thread a needle. For indoor or night shots, there was a flash unit that synced to the camera and fired flash bulbs.

After taking the pictures, the film was developed like any other roll of slide film. Instead of being cut up and mounted into individual slides, the film would be left uncut as one long strip.

With other 3D cameras, you'd have to carefully measure and cut the left and right images from the filmstrip, then manually trim and mount them into a frame, spacing and aligning the images by hand. One slip of the razor blade or sloppy alignment and you'd have ruined your picture. Again, too much trouble for most casual photographers.

empty View-Master reel. No tricky alignment needed: the precise slots snugly held the film in perfect registration. There were even blank spaces on the reel to write your own captions for each of your seven images.

Because the View-Master reels you made would fit any View-Master viewer, you could send reels to friends and they could view them with the viewers they already had or could easily get. But for the ultimate in sharing your 3D View-Master pictures with a group, there was the Stereo-Matic 500 projector, the last part in the View-Master system.

No worries. The next element of Gruber's View-Master system, The View-Master Film Cutter, made it easy to create your own V-M reels. It had twin precision cutting dies and a rock-solid filmstrip advance mechanism. With a twist of the knob, the sprocketed film automatically locked into position. You pressed the lever down—*kaCHUNK!*—and both left and right images were crisply punched out with a single stroke. You inserted each chip of film into its matching slot in an

The Stereo-Matic 500 projector featured twin lenses that automatically aligned the left and right views as you focused. Its polarized filters matched the lenses in the special glasses the audience wore. Your photos were projected in bright, colorful, thrilling 3D for all to see up on the silver screen (a metallic screen surface was required to maintain the polarization).

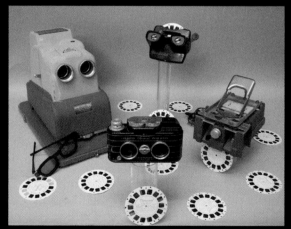

You can see this picture in 3D by "free viewing." Hold the page at arm's length and cross your eyes. Relax your focus and adjust your gaze so that you see three boxes. The center overlapped image will "pop" into 3D as you focus on the black camera in the middle.

VIEW THIS ANAGLYPH IMAGE USING RED/GREEN FILTER 3D GLASSES. REMEMBER "RIGHT EYE = RED."

The entire View-Master 3D system worked quite well. With rugged construction and clever design, my camera, film cutter, and projector still work flawlessly after 50 years. I've taken thousands of 3D images: wedding albums, birthdays and holidays, baby pictures and family events, scenic travel photos, even some "artsy" shots—all with great results.

Traveling with the View-Master camera is great fun and a real conversation starter among fellow travelers and other photographers. The camera's unique sliding "guillotine" shutter makes a distinctive *pishhhhhht-click* sound with each picture you take and gets plenty of attention, wanted or not. I've often gotten curious looks and been asked about my unusual camera on trips abroad. *Qu'est-ce que c'est cette caméra? ¿Qué clase de cámara es ésa?* I'd just pantomime holding up an imaginary View-Master viewer while make the "flicking the lever" gesture and I'd get knowing smiles and nods back.

Once an all-ages, all-family product, today the View-Master brand continues on as a preschool toy. Making your own 3D View-Master reels is still possible, though you'll have to be resourceful. Vintage View-Master cameras, projectors, and accessories can be found on eBay (and at collector prices!). Kodachrome film is now just a colorful song lyric, but Fujichrome 35mm slide film and developing is still available.

One way or another you, too, can still make friends and family go "Wow" with your own personalized View-Master 3D photos. Determined do-it-yourselfers can still have fun making their own View-Master 3D reels using modern digital cameras. Service bureaus can print your digital files to high-quality transparency film. Designer Shab Levy offers an ingenious system and kit for View-Master reel making (see below).

You can also hack a low-cost hardware store item into a slide bar for taking stereo images with your regular digital camera or smartphone.

Level Best 3D Camera Slide Bar Hack

Materials and Tools

Inexpensive T-square level
Two ¼"-20 bolt and nuts
⁵⁄₁₆" drill bit and drill
Camera and tripod

For taking 3D photos with a regular camera you need a tripod and stereo photo slide bar. Professional slide bars cost $50 or more, but you can make a DIY slide bar fast and cheap that will work just as well. Here's how:

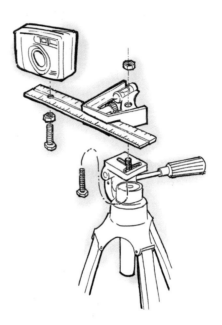

Find a T-square with a frame that's big enough to sit flat on the top of your tripod head. Drill a ⁵⁄₁₆" inch hole in body of the level to fit onto the tripod's stud, or if needed, add a ¼"-20 bolt and nut to attach to the tripod.

Drill another hole near the end of the steel rule to hold your camera. Attach your camera with a ¼"-20 bolt and a nut. Carefully thread the bolt into the camera's socket, and then tighten up the nut behind the rule to secure the camera. You're done! (If you have an adapter that will mount a smartphone to a tripod, use that to mount to the slide bar.)

To shoot 3D, level the camera on the tripod. Loosen the T-square thumbscrew and slide the bar so that the camera is against the tripod head. Retighten the screw and take one picture. Loosen the screw, slide the camera bar over 2½″ (that's the standard intraocular spacing for normal 3D picture taking) and retighten the screw. Take your second picture. Now you have a stereo pair of images, one for each eye. Of course, if your subject moved between the two exposures the 3D effect is lost. Works best with static scenes.

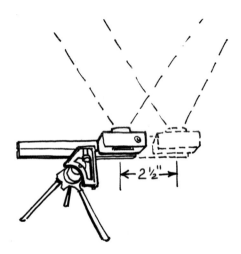

You can print out your pictures side-by-side for free viewing (like the 3D pictures in this chapter), or use Photoshop to combine the images into a red/green anaglyph and view with some red/green glasses. To make your own View-Master reels, I refer you to Shab Levy's slick kit with everything you need: precision die-cut empty reels, software templates, a mounting rig and complete instructions. As of this writing, you can find it at: gravitram.com/a_kit_for_making_your_own_custom.htm.

Colorful Comic Kaleidoscope

Ordinary kaleidoscopes need ambient light to work: you view by pointing it to the light. This one has built-in illumination so you can hold it right up against a printed page, like a comic. The special RGB LED (RadioShack 276-016) has a color-changing feature that continuously cycles through a rainbow of colors, making viewing black-and-white comics a trippy treat!

Make the body from a 1¼″ diameter Tube-Pak cut to 6½″ length. Cut three strips of black ABS 1⅛″ wide and 6½″ long. Arrange in a triangular shape with the shiny sides in and insert into the tube. The three shiny surfaces make a great first-surface mirror kaleidoscope!

Materials and Tools

1¼″ diameter PETG Tube-Pak tubing	2032 3V button cell battery
2 vinyl caps for above	Battery clip for above
Black ABS, enough for three strips, about 6″ x 8″	Compass
Tricolor RGB LED (RadioShack 276-016)	Hobby knife
Miniature SPST momentary contact push-button switch	Solder iron and solder
Hook-up wire	¾″ and ¹⁄₁₆″ drill bit and drill

Cut a 1⅛″ diameter disc out of ABS. Drill or grind a ¾″ hole in the center. Drill three small holes near the edge and thread the LED's legs through. Bend the legs to ensure the LED will shine inside of the mirrors when the disc is in place. Attach the switch to the disc with superglue.

Wire the LED's flat-side lead to ground of a 3V button cell battery holder and wire the center leg to positive. Then wire an SPST momentary switch between the third LED leg and ground. Press the button to turn on the LED and then cycle through its modes, the last of which is a cycle that goes continuously through all the colors.

Cut a 1³⁄₁₆″ hole in the center of both caps. Make a hole for the switch button the side of one cap and make a matching notch in the end of the tube. Position the disc with the LED in place inside and slip the battery between a mirror and the tube. Slide the cap on with the switch through the small hole and in the notch. Put the other cap on the other end. If needed, cover the back and side of the LED with a small piece of black electrical tape to shade your eye from glare.

Slide the scope directly over your favorite comics and press the button to select a color mode—cool!

Scan to see the Colorful Comic Kaleidoscope in action!

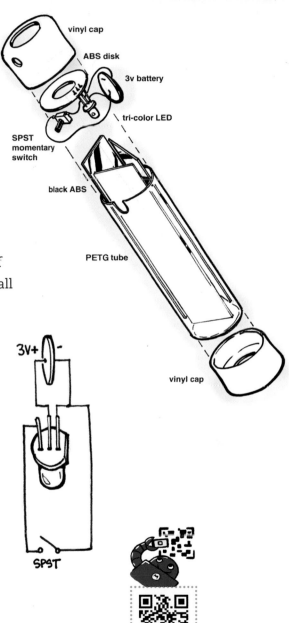

Chapter 4

Thermoforming Plastics

The 1960s were a golden age for toys in America due to a timely combination of postwar factors. The Baby Boomers were in their prime toy-buying years ("for ages 6–12"), while the boom in the economy created more purchasing power for their parents. For the first time, toys were sold using network television advertising campaigns, launching nationwide fads—fast, fast, FAST! The graduate of the 60s was told there was a "great future in plastics," while at the same time the space race spurred an interest in science education and technological toys.

Productive Plastic Playthings
A Look Back at 1960s Maker Toys

Among the time-tested playthings like dolls and trucks came a new category of toys: "make-and-play." There had been creative kid crafts before, like paint-by-number kits or *Erector* sets, but these modern maker toys inspired a generation of kids to mass-produce their own creations using miniature, at-home versions of industrial manufacturing methods and advanced "space-age" materials. With clever research and development and bold marketing approaches, Mattel led the way in make-and-play.

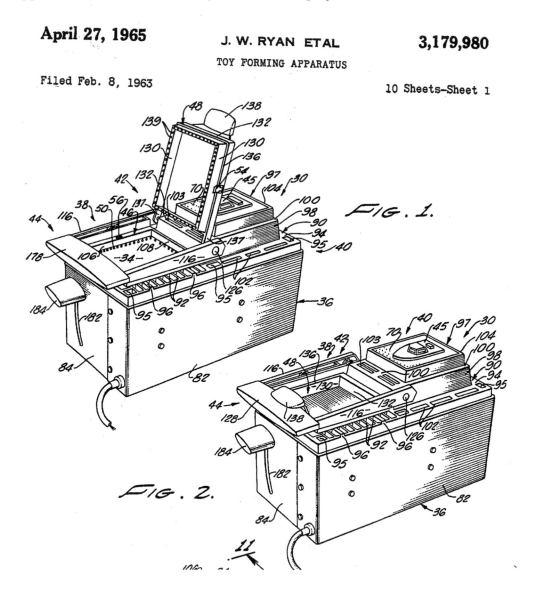

April 27, 1965   J. W. RYAN ETAL   3,179,980
TOY FORMING APPARATUS

Filed Feb. 8, 1963   10 Sheets—Sheet 1

FIG. 1.

FIG. 2.

→ The Vac·U·Form in Action—in Five Steps ←

1. Choose a mold and place on the vacuum stage.

2. Load a sheet of thin plastic into the frame and place it over the heater.

3. When the plastic is soft, swing the frame over the mold and pump the handle.

4. When cool, remove the plastic.

5. The finished vacuum form! Trim away the extra for the finished part.

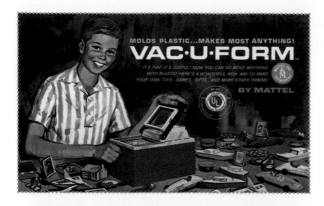

My favorite childhood toy was the Mattel *Vac·U·Form* machine. The pungent smell of melting plastic filled my bedroom as I spent many fun hours molding little cars, bugs, toys, and signs. The resultant flimsy toys really weren't much fun to play with, but that didn't matter. The way the flat plastic instantly changed shape due to an invisible vacuum was magical.

Invented by Eddy Goldfarb, Mattel's 1963 *Vac·U·Form* used air pressure to mold small squares of colorful polystyrene into three-dimensional shapes. After three minutes on the 110-volt plug-in heater, the softened plastic was stretched and formed over a mold with a few quick strokes of the pump handle. Vacuum forming thermoplastics, the process used by Boeing to make jet airline interiors, was now at the fingertips of kids across America.

And those fingertips often got blisters! Unlike today's super-safe car-seated and bike-helmeted kids, Baby Boomers braved the dangers of toys like the *Vac·U·Form*. Its exposed heating plate reached the skin-sizzling temperature of 350°F. You can tell it's Mattel, it's . . . OWWWWW! No matter. With the awesome power of the *Vac·U·Form* pressurized plastic, kids cranked out cars, planes, signs, disguises—all kinds of mini toys using the many molds that came with the toy.

Just as innovative as the *Vac·U·Form*'s design was its marketing. Like Barbie with her never-ending array of clothes (sold separately!), the line of *Vac·U·Form* toys had dozens of accessories and refill kits for making jewelry, medals, badges, airplanes, animals, boats, military vehicles, and more. Each kit supplied more molds, more projects, and more sheets of plastic, including both transparent and metalized chrome in gleaming colors. A nation of kids was mesmerized by the *Vac·U·Form*'s TV commercial featuring the magical moment of transformation: molded shapes took form before your eyes, set to the peppy ba-boom beat of timpani.

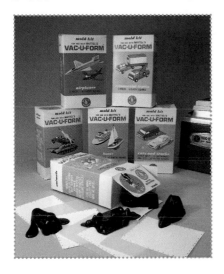

A familiar smell can trigger a flood of childhood memories, and just one pungent whiff of burning plastic is enough to evoke for me the *Thingmaker*, a spin-off of the *Vac·U·Form*. The same heater that softened stiff sheets of styrene could be used to cure liquid plastisol. This goopy mix of polyvinyl-chloride in a solution of plasticizers is used to manufacture soft parts like tool grips, squeezable coin purses, and flexible fishing lures. Renamed *Plastigoop* and packaged in handy squeeze bottles, the protean plastic came in a dozen colors, including an exotic glow-in-the-dark formula. The *Thingmaker*'s molds were made from die-cast zinc, the same metal used in toy cap guns, which reproduced each tiny mold detail, from the hairs on a spider to the gruesome stitches on a shrunken head.

STOCK NO. 4523

Like its older brother the *Vac·U·Form*, the *Thingmaker* offered the excitement and danger of high-temperature techniques. The fun began by filling the mold with various colors of Plastigoop. Heating the mold on the *Thingmaker*'s oven cured the Plastigoop into a wiggly gel. The last step was to quench the finished mold in a pan of water, making a satisfying blast of steam. The only concession to safety was the wobbly wire handle used to lift the smoking hot molds.

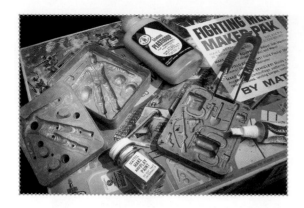

There were many different themes for *Thingmaker* sets, but the most successful was Creepy Crawlers. The TV commercial featured a James Mason sound-alike voice-over, drolly describing how to scare Mom and annoy newspaper-reading Dad with the wiggly worms and rubbery bugs. Mattel went on to produce Giant Creepy Crawlers (with flocking to make the bugs fuzzy), Creeple People (cute/ugly trolls with interchangeable heads, arms, and legs that connected together on a pencil), Fighting Men (army men with wires for bendable arms), and Fright Factory (shrunken heads and skeletons).

For kids who wanted to stuff that was less scary or gross, there were sunnier-themed sets: Fun Flowers, Mini-Dragons, Picadoos (plastisol pixel pictures), Zoofie-Goofies (pipe cleaner animals), and even sets that made rubbery versions of comic strips like Charles Schulz's Peanuts. All required, of course, ever more bottles of colorful Plasti-goop, sold separately.

Other toy companies sold similar items. Topper Toys produced a thinly veiled

knock-off of the *Thingmaker* called *Johnny Toy-maker*, a casting toy that made racing cars for boys and rose-decorated sunglasses for girls. Kenner's 1964 *Electric Mold Master* boasted that it "makes *solid*—not vacuum-formed toys." Boys could mold army men and mini toy guns that shoot, and girls created small doll and accessories, including multicolored "miniature teenagers at a birthday party" with tiny bongo drums, record players, and soda glasses.

Perhaps the weirdest heat-powered play-thing was the *Strange Change Time Machine*. Instead of merely molding raw plastic into mini toys, this clever contraption provided endless play with 16 different time capsule creatures. "Create 'em! Crush 'em! Create 'em! Again and again—in the Strange Change Machine!" The time-lapse TV commercial showed a square plastic lozenge magically melt and transform into an octopus. Reheated, the mini creatures were loaded into a vise-like press and were squeezed back into the original square shape complete with an embossed Mattel logo. You didn't really "make" anything, but it was a mystifying trick!

A space age "shape memory" plastic made it all possible. During manufacture the molded plastic creatures were radiated with a high-energy electron beam that cross-linked the polymer's molecules, permanently locking them into shape. These extra connections within the shape could be deformed temporarily into a square brick, but when reheated they sprang back into their original creature shape.

A STRANGE CHANGE TOY
FEATURING THE LOST WORLD
16 FANTASTIC TIME CAPSULES!
CREATE 'EM!
CRUSH 'EM!
CREATE 'EM!
AGAIN AND AGAIN—
IN THE TIME MACHINE!
CREATE THIS—FROM THIS ▶

The look of the toy was pure 1960s with snazzy, metallic red housing, shiny zinc fittings, and a transparent transformation chamber with a swiveling door. The assortment of creatures included tiny dinosaurs, kooky spacemen, and mini monsters. The instruction sheet flipped over to make a jungle island backdrop. Even the package's plastic shipping tray was vacuum formed into a sculpted volcanic rock pit. The entire toy was like a sci-fi monster movie set shrunk down into miniature toy form.

The name *Time Machine* was an apt one: it felt like you needed one. It seemed to take *forever* for the toy's oven chamber to warm up, and even longer for each of the creature shapes to unmold and then cool down. The ominous warning stamped on the top of the toy, CAUTION—CONTACT WITH RIVETS OR PLASTIC PARTS MAY CAUSE BURNS, was also from another time, one before the Child Safety Protection Act.

One hugely popular 1960s maker toy did have an infamous safety problem. Like the *Thingmaker* before it, *Incredible Edibles* used a heated oven (with a hinged cover in the form of a bewigged, buck-toothed bug) to let kids mold squiggly spiders and squirmy worms. The fresh twist was that the finished product was actually edible. The ingredients listed glycerol and tapioca starch sweetened with sodium cyclamate and saccharin.

The moldable comestible was dubbed Gobble-Degoop and marketed as "Sugarless!" Parents were repelled more by the taste than the fun "gross out" theme, but kids gobbled it up. The unforeseen problem was that diabetic kids were sickened by the mixture's starch, which turned into sugar when digested. After $50 million in sales, the FDA allowed Mattel to put warning stickers on all the toys already out in the stores instead of recalling them.

An earlier candy-making toy promised real sugar in its most kid-appealing form: cotton candy! Commercial cotton candy machines were expensive and complicated with spinning electrical coils and strong motors. A kid's only chance for the rare treat was a trip to the circus or state fair. But in 1962 Hasbro's affordable, battery-powered *Hokey Pokey Cotton Candy Maker* spun real cotton candy at home anytime.

The patented design had a red scalloped plastic motorized base with six ball-shaped feet that supported a deep aluminum drum. Mom loaded a few spoonsful of sugar into a metal cup and heated the cup over a kitchen burner. Using the metal tongs, she lifted the cup of molten sugar into place atop the toy's central shaft and turned on the motor. The molten sugar was flung out of the madly spinning cup in an instant to make one small batch of cotton candy. The metal cup then could be cooled down on the included asbestos stand. Asbestos? Flying molten sugar? The moon-faced kids illustrated on the package happily twirled their paper cones of cotton candy, sweetly oblivious to any potential domestic dangers.

Toys offer kids a way to emulate grownups and play pretend, and maker toys are no different. Dad's basement woodworking "man cave" was the inspiration for Kenner's 1962 *Motorized Home Workshop*. Although scaled down, the versatile styrene toy version could be configured into seven different power tools, including lathe, jigsaw, drill, and sander, just like Dad's Shopsmith.

The gimmick that made it all safe for Junior? Instead of using wood, the raw material was colorful, open-cell urethane foam, and the saw and drills were made of plastic. The battery-powered motor "stopped at the press of a finger." It was fun to make clouds of foam "sawdust" with the motorized tools as you cut, sanded, and turned your project on the lathe. The box blurb promised "planes that fly and boats that float," but the flakey foam made fuzzy and feeble results. Oh well, the fun was in the making.

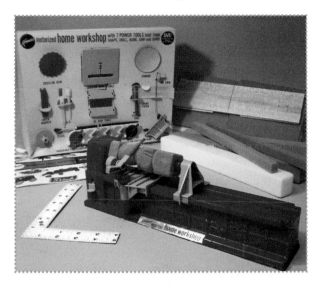

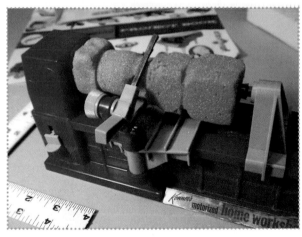

When it comes to mass-producing plastic parts, the manufacturing process used by toy companies is injection molding. A two-part metal mold is held tightly closed in a horizontal hydraulic press as molten thermoplastic is injected at high pressure into the mold cavity. The mold is opened and the cooled plastic part is ejected. Unlike vacuum forming or casting, injection molding with its closed mold creates parts with perfect detail on all sides.

What better maker toy than a miniature version of the very process used to make real toys? Mattel's *Injector* from 1969 had a plug-in heater and a hand-powered injector piston.

To use, you just slid open the chamber and inserted a small plastic pellet (called Plastix). While the plastic melted, you chose your mold. A small toggle provided the clamping pressure to hold the mold halves together. When you slipped the mold under the nozzle and pushed down firmly on the lever, it would squirt a piston full of plastic into the mold. With different themed versions you could mold your own Hot Wheel Factory car bodies or Western World cowboys and Indians.

→Toy Maker Mascots←

Depicted on the packages and molded right into the toys were the toy company mascots. These peppy pitchmen delivered TV taglines and were even pressed into service enlivening otherwise bland instruction sheets. Matty Mattel gave tips on vacuum forming, Hasbro shilled the latest Hassenfeld Brothers offerings, and the Kenner Gooney Bird screeched "It's Kenner! It's FUN—SQUAWK!"

So where are the maker toys of today? Social trends have changed and so have playthings. Today's over-scheduled kids don't have as much free time to spend waiting for long heating and cooling cycle times to make a leisurely batch of Creepy Crawlers. Kids now "grow older younger;" whereas toy manufacturers of yore could market toys to kids ages 6–12, today's 10-year-old is already using computers and getting her first iPhone. Tweens aspire to create with real, adult materials and tools: they're at the yarn shop, art store, or Maker Faire, not in the toy store. They can try their hand with sophisticated 3D printers and laser cutters found at maker-spaces and schools.

Toy makers must now focus on younger ages that are inherently less patient, capable, or interested. Despite clever updates by manufacturers to address modern toy safety standards, it's hard to compete with the high-tech appeal of video games and electronics.

The economy has changed and toys reflect those trends. Just as manufacturing industries are being replaced by an information-driven service economy, early maker toys with their "thing-making" play pattern have given way to the virtual experiences of electronic games, apps, and websites aimed at kids.

Still, the spirit of DIY lives on and these early maker toys engendered an interest and curiosity in the very kids that grew up to be many of the Makers of today. Despite the huge popularity of "just assemble" activity toys like LEGO, it's encouraging to see young kids together with their parents learning to solder and building real hands-on creative skills at Maker Faires.

PHOTO COURTESY TROY FISCHER

Kitchen Floor Vacuum Former

From the lid on a Starbucks coffee, to the curved interior panels of an airliner, vacuum-formed plastic parts are everywhere. And for good reason: vacuum forming makes light, durable, and cool-looking 3D parts. Here's how to cook up plastic parts quickly in your own kitchen with a low-cost, maker-friendly rig.

Large commercial machines have built-in vacuum pumps, adjustable plastic-holding frames, overhead radiant heaters, and pneumatically raised and lowered platens. The Kitchen Floor Vacuum Former is much simpler. It uses the broiler in your kitchen oven to melt the plastic. You manually flip the plastic onto the form. A household Shop-Vac supplies the suction. All you have to build is a simple wooden frame and a hollow box.

→ Operating in a Vacuum ←

Vacuum forming is a simple two-stage process. First, a sheet of thermoplastic material is softened with heat. Then, using suction, the pliable plastic is pulled and stretched over a form. When the plastic cools, it stays in the new shape. Vacuum-formed parts are uniform and stackable, which makes the process ideal for mass-producing things like blister cars, coffee cup lids, cookie package trays, egg cartons, and costume masks. And unlike injection molding, the tooling needed to vacuum form parts is very low-cost, low-tech, and easy to make!

Materials

2 × 4 lumber (1½″ × 3½″) I used fir, which is cheap, easy to staple into, and being wood, reasonably safe to handle when hot. You just need four short pieces, probably 2 feet or less each, so scraps are fine.

Plywood to make a shallow box. I had some scrap ½″ plywood, but you can use nearly any material you happen to have: plywood, particleboard, framing lumber, whatever. You could even repurpose an old dresser drawer or a deep picture frame. Get creative!

Pegboard—2′ × 2′ should be enough.

Short piece of dowel or 2 × 2 lumber, or other small wood scrap

Floor nozzle attachment for vacuum cleaner

Drywall screws

Duct tape

Plastic sheets—polystyrene works well, but you can use nearly any thermoplastic material. 0.030″ is a good thickness, or try thicker if you need a stronger part.

Shop-Vac or other vacuum cleaner

Oven mitts

Surface gauge (optional)

Staple gun and staples

Tape measure

Saw

Metal straightedge

Drill bits and drill

Heat gun

Plastic scribe (not shown)

Form material—I used rigid urethane foam, but you can use almost anything.

Oven with broiler

Woodcarving tools and a sharp knife

Sandpaper

Screwdriver

A few coins or window screen

1. Make the Frame

1. Measure the interior of your oven, then subtract a few inches from the width and depth for clearance. This gives the size of the biggest frame you can use, which in turn determines the maximum size sheet of plastic you'll be able to mold. My oven is 21″ × 16″, so I made an 18″ × 13″ frame.

2. Measure and cut lengths of 2 × 4 you need to make your frame, allowing for the thickness of the sides. Make square cuts with your favorite saw: circular, jig, or handsaw. My 18″ × 13″ frame calls for two 18″ pieces and two 10″ pieces.

3. Assemble the frame using 2 screws in each corner for maximum strength. Stagger the fasteners so as not to split the wood.

2. Make the Vacuum Box

4. Calculate the dimensions of a shallow box with a top slightly larger than your frame. For my 18″ × 13″ frame I made a 20″ × 15″ box.

The height of the box is not critical. I made the height of the box just large enough to mount the vacuum floor nozzle along one side.

When you have the dimensions, cut four sidepieces and a bottom out of plywood, accounting for the material thickness again.

5. Cut a top panel for the box out of pegboard, large enough to overlap all four sides. The pegboard's holes are used to suck air from under and around the form.

6. Drill a large hole in the side of the box where you will mount the vacuum floor tool. This is the main air vent.

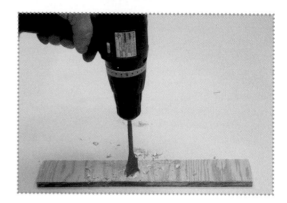

7. Use screws to assemble the four sides, and then add the back.

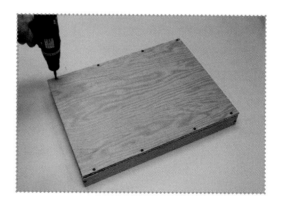

8. Cut a piece of wood dowel or 2″ × 2″ to make a post that will support the center of the top panel. When vacuum forming, the top panel may tend to bow in. This support post prevents that.

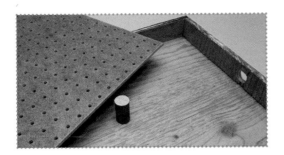

9. Screw the post down from the back of the box, and then fasten the pegboard to the sides with screws.

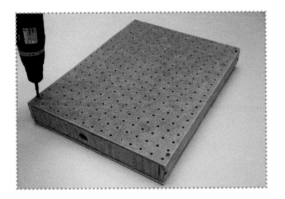

10. Mount the vacuum over the side hole with small screws.

11. Finally seal all the edges with duct tape. Don't forget the seams around the vacuum tool. Tape a border around the top, but leave an open area in the center slightly smaller than your frame. Your Kitchen Floor Vacuum Former is ready to use!

➔If You Don't Want to DIY←

If you don't want to make your own vacuum former, or you just prefer to stay out of the kitchen, there are a couple commercially made units that might work for you.

The Nichols Therm-O-Vac is a completely portable design with a fold-out heater, flip-over clamp frame and a built-in vacuum pump. It has clever adapter plates so that you can use small pieces of plastic to fit various sizes of molds—no waste. I love mine! (It was even featured on the silver screen in the 1986 movie *F/X* as it molded a life mask.)

If you only have to make very small parts, maybe Micro-Mark's aptly named Compact Vacuum Forming Machine will fit the bill. Small in price, too.

See the appendix for more details on sources.

Tiki Mask

Here's an idea for a fun, first vacuum forming project: a kooky tiki mask. It's easy because there is only one part to make. There's plenty of leeway in tiki design: almost any approach will work. Make it simple or complex, symmetrical or free form, brutal or refined, friendly or freaky—it's totally up to you. A mask is also a great example of what vacuum-forming does well: making light-weight yet strong 3D parts quickly and easily.

1. Form Materials

You can make your form out of nearly anything. Wood is a good choice for most simple forms: it's cheap; easy to drill, saw, and sand; and strong enough to make many parts. An especially good wood to use is jelutong. Its soft grain-free texture makes it a pleasure to carve and sand.

You can also make forms out of plaster. Sculpt a shape out of clay, then cast liquid plaster all around it. After the plaster hardens, remove the clay and drill air venting holes through the bottommost features of the mold. Sucking plastic down into a concave "female" mold cavity will give parts with a very detailed surface that will match your clay original perfectly.

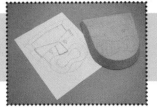

If needed, combine different materials to build up your form. Found objects can be a great source of shapes and details that would be hard to make from scratch. You can add small plastic letters, caps and lids, or pieces of toys to build up your form. I keep a collection of basic shapes that I can combine and reuse to make new forms.

I used urethane foam for my mask. The surface of the mask doesn't need to be perfectly smooth and the foam is especially easy to cut and carve with woodworking tools. You can order urethane foam in various densities for more durable forms, but if you're just making a few parts you can use florist's foam (get it at a hobby store). If you need just one rough-surfaced part you can even use leftover Styrofoam. It's not very strong and it may melt and fuse a bit to the hot plastic sheet when being formed, but I've used it successfully on occasion.

To form the foam, cut out the basic outline with a saw, then drill or cut out smaller sections as needed. File or sand to create curved surfaces. Use a rotary tool to sand and grind out smaller shapes. Sculptor's tools or dental tools are great for creating small details or textures. You can crunch the foam to make small depressions.

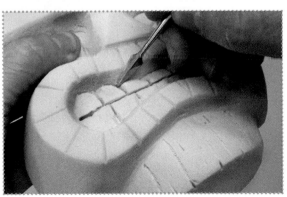

2. Mold Design

Here are some things to keep in mind when making a vacuum form:

Undercuts Avoid undercuts in your mold shape and in any details. The softened plastic sheet will stretch and wrap around underneath all the features of your form. After the plastic cools and gets firm it will be impossible to get the trapped form out.

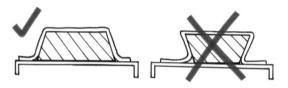

Draft angle To make it easier to get the plastic off your form, avoid using vertical sides. Make your form so that there are a few degrees of angle (draft) on all sides.

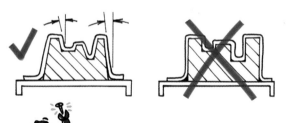

Vent holes For concave shapes or details, drill a series of tiny vent holes. These holes will let the air flow around your mold and suck the softened plastic down into your form.

Extra height If you plan on trimming your part away from the rest of the sheet, add a little extra height to the form. That will ensure that your final part won't have any unformed edges.

Maker Tip

A very useful tool to have is an extra-long 1/16" drill. The flutes are only in the very tip of the drill's extra-long shaft, so you have to be careful. Drill only partway, withdraw the drill to free the chips, then re-insert and drill some more. Repeat until you've drilled completely through.

3. Plastics and Sources

You can use nearly any thermoplastic material. Polystyrene sheet works especially well and is low-cost. It comes in large 4′ × 8′ sheets in various thicknesses. Check with your local plastic supply company; they may have a bin of odd sizes, cut-offs, and leftovers that may suit your project. 0.030″ is a good thickness, but you might want to try something thicker if you need a stronger part. IASCO–TESCO (iasco-tesco.com) sells small sheets of styrene in various thicknesses and in solid colors.

4. Vacuum Forming

When you're ready to vacuum form, get set up on the kitchen floor near your oven. Hook up the Shop-Vac and test.

Place your mold on the top of the box. To help air flow, place your mold on top of some coins, which will act as spacers. You can also use a small piece of window screen.

Measure the plastic to fit your frame and score with a sharp knife. Bend the sheet backward to snap at the score line.

Place the plastic over the frame and staple all around. Don't skimp on the staples: the softened plastic will try to pull away so staple every inch or so.

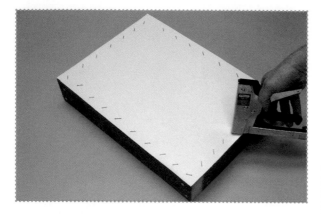

Place the frame, plastic side up, in the oven four or five inches away from the broiler. Have your oven mitts ready and set the broiler on high.

BE CAREFUL! Watch the plastic carefully as it warms! First it will look wavy, then it relaxes back flat, and finally the plastic will begin to sag in the center. Test the plastic by pressing down near a corner with a pencil eraser or chopstick: if it's soft, it's ready to remove. If left too long under the heating element, the plastic will smoke and burn!

With your oven mitts on, remove the frame and flip it plastic side down. Quickly and carefully, place it over the form and press down. When the frame is pressed firmly flat against the top of the box, turn on the vacuum.

Whoosh! The plastic is instantly sucked down over the form by vacuum power!

If your part has some intricate detail or hard to mold feature, use a heat gun to soften the plastic in that area.

BE CAREFUL! You could melt right through the plastic!

Continue pressing down on the frame for 20 more seconds while leaving the Shop-Vac on until the plastic has cooled. Then turn it off and lift up the frame. Your form may fall out, but typically it will stay inside the formed plastic. To remove the form, press on the back gently or tap the frame firmly on a table or counter so that the form can fall freely. You may have to carefully pry the form out a little at a time with a screwdriver or knife.

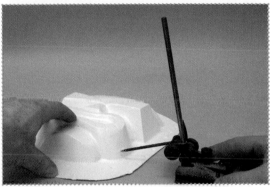

5. Trim and Finish

Carefully remove the plastic from the frame. Look out for sharp staples!

Score the plastic and snap to remove the formed part from the rest of the plastic. Use a surface gauge to scribe a mark all around the part for a more precise trim.

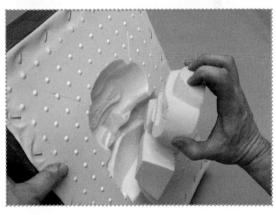

If you need a lip on your part, trace around the shape leaving a flat border all around. Score with an X-ACTO blade and snap. Clean up the edges of the plastic with sandpaper. Drill, trim, or paint your part for your final application. I'm finishing this one as a mask with large eyeholes, small side

holes for elastic, and a rim around the bottom for extra strength.

Here's the finished tiki mask. I spray painted the entire mask blue, then added shading by spray painting from the side only with purple accent color. Then I hand-painted the small details for the flowers, teeth, and eyes. To wear the mask, add some elastic cord tied at each end with a knot.

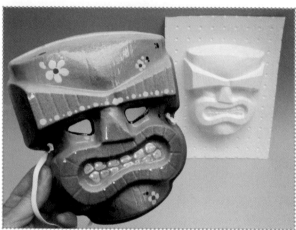

Glowing Tiki Lamps

Take your tiki mask to the next level by adding lights to make a glowing lamp.

If you made your mask with a lip around it, place it on another piece of styrene sheet and trace around it. Score on the line with a hobby knife and snap off to make a back panel. Wire up a low-voltage light using a small battery pack (look for the kind with the built-in on/off switch) and an incandescent light bulb. Hot-glue the battery pack and light bulb to the back panel so that the bulb will illuminate the panel but not be seen through the mask's eyeholes.

Use the Velcro dots around the rim to fasten the mask to the back panel. You can easily remove the mask to turn it on or off and change the batteries.

Paint the dowel to look like a piece of bamboo. Add some painted lines and spots as well as hitting it with some random patches of black spray paint (see picture).

Hot-glue the back pane to the dowel. Stick it in the ground or place it on your deck for your next backyard party!

Materials

Tiki mask
Extra styrene sheet
1″ wood dowel in the length you want
AA battery pack
Wire
Light bulb
Velcro dots
Hot-glue gun
Soldering iron
Hobby knife
Paint

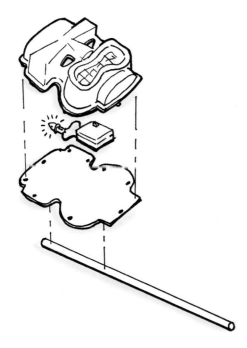

Tiki Jell-O Mold

Materials

Tiki vacuum form
PETE sheet
Flavored gelatin and water

Here's another idea for your backyard tiki party: a tiki-shaped Jell-O mold!

For this, I made a vacuum form of the tiki mask but instead of using styrene I used polyethylene terephthalate. Sometimes called PETG or PETE, it's used to make food and beverage containers since it's transparent and tough and not as brittle as styrene. Look for the 1-in-a-triangle plastic recycling symbol molded in your fast-food or take-out container, or food packaging—that's PETE! Find sheets in various thicknesses at your local plastic supplier or online.

I used it here to make a mold for casting gelatin. Just vacuum form as before, but this time leave a generous rim all around the mask and don't cut any eye or strap holes.

For best results, make the gelatin double strength (or even more for a chewy treat). Add 2½ cups of boiling water to two packages of gelatin. Stir until dissolved. Pour into the mold and refrigerate for three hours until firm. To unmold, dip the bottom of the mold into a pan of warm water for 20 seconds. Place a serving plate upside down over the mold and flip both over.

Maker Tip

What else can you make with your vacuum form tiki head? How about a giant tiki ice sculpture: just freeze water and stand it up on your luau buffet or float in your punch bowl. Cast chocolate in custom shapes or make personalized birthday cake toppers! Or take your mold to the beach and make sand sculptures or to the mountains for making snow shapes.

Marble Maze Games

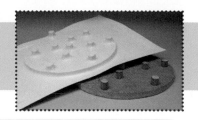

Here's a twist on a classic skill game: the tip-and-tilt marble maze. Hold the maze tray with two hands and tip and tilt to roll your marble through the maze to the goal. Make it, play it—then hack your maze to create different games. In a multiplayer version it's a race: the first player to get his marble through the maze and to the finish wins!

Materials

Styrene sheet for vacuum forming
12″ × 12″ ½″-thick plywood or particleboard
¾″ wood dowel
Marbles
Plastic tubing (I used thin-walled, blow-molded, recycled tubing from a respirator, but any tubing big enough for your marble will work.)
Zip ties

Tools

Kitchen Floor Vacuum Former
Hobby knife
Pencil
Drill and bits
Wood screws

1. Make the Form

Make a form by cutting an 11″ disk out of ½″ plywood or particleboard. Cut the ¾″ wood dowel into ¾″-long pegs. Arrange the pegs on the disk to create a pattern of holes for a maze obstacle course. Each vertical peg on the male form will create a marble pit on the inverted vacuum-formed female part. Mark the positions of the pegs on the disk and drill a small hole for each. Use a wood screw through the backside of the disk to attach each of the pegs. Later, if you like you can quickly reposition the pegs to make a completely different maze layout.

2. Vacuum Form

Use your Kitchen Floor Vacuum Former to make the maze tray. When cool, remove the form and trim it to make a rimmed tray. While you're at it, make a couple more trays.

Cut off the bottom of one pit to make a "finish" hole.

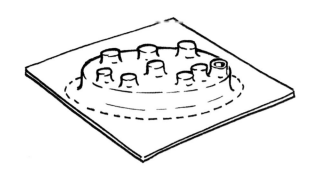

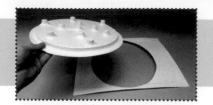
3. Fabricate the Maze

Measure, score, and snap off some ½" wide strips of styrene. Hold a strip in position to make a wall for your maze and mark the length to suit. Score at that length and snap off. Hold the strip in position on the maze tray and glue with some MEK solvent (see the appendix for sources). Attach more lengths of plastic strips together to create a path of maze walls around the pits. Test it: tilt and tip the tray to get your marble through the maze, around the pits, and down the hole. Too hard? Reposition the walls until it's just challenging enough! You can make periodic "safe" areas along your maze: little corner pockets on either side of a pit that you can tip your marble into and then rest before attempting the next part of the maze. Draw a dashed line to show the final path through your maze.

➜ Fascinating Fascination ⬅

Remco's *Fascination* is one of the original 1960s marble maze games with a simple but a fun gimmick. Each player has a handheld maze and three marbles to maneuver from the start to the finish. The first player to get all three marbles in their finish holes lights their color-coded light! How did it work? Super simple: the marbles were steel ball bearings and each finish hole had two brass contacts on either side, all wired up together in series with a battery and a light bulb. Getting the three conductive balls in the three holes completed the circuit and lit the bulb—a winner! Does this give you any ideas for making your own update? Add mechanical or electronic sounds? Arduino logic functions to trigger other electro-mechanical outputs? Game timers? Vibrating motors for haptic feedback or moving hazards? Add a Raspberry Pi to connect your game to the World Wide Web? I leave it to you . . .

4. Hack the Design to Make New Games

Race against the clock Use a timer to see how quickly you can do the maze. Fall into a pit? You must start over from the start—hurry, the clock is running. You could even hack an electronic timer's stop button with an extra switch, triggered by the marble coming down through the finish hole.

High score maze Mark each hazard pit along the path with ascending point values—the further you get, the higher your score.

Multi-marble Use more than one marble at a time. Ulp! How many marbles can you get all the way to the finish?

Color puzzle maze Mark the pits with different colors—first to get all the colored marbles into their matching pits wins!

3D maze Make a multilevel stacked maze! Drill small holes in the rim of each maze and use screws to fasten 4"-long lengths of wood dowels to stack the mazes. Start your marble on the top maze. The marble falls down to the next level as you go. Can you go all the way?

Multiplayer maze For multiplayer marble maze action, connect the trays with long flexible tubes (I scrounged air tubes from a hospital ventilator). Connect one end of the tube to the finish hole of one maze, then zip tie the other end to the rim of the other maze. Get your marble through the maze, down the hole, and then lift your maze to send the marble through the tube and out into the other maze for the next player.

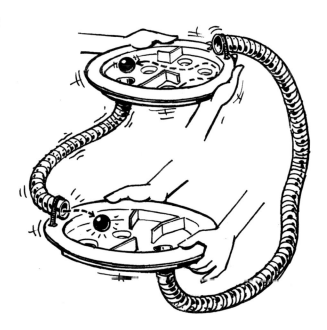

Connect more mazes in series to make a giant multiplayer relay maze! Or connect two mazes with two tubes for back-and-forth action. Each player gets three marbles: first one to get all his marbles into his opponent's maze wins!

Strip Heater

This may be the simplest project in the book, yet one of the most versatile. The strip heater is an electrically heated coil that is stretched in a straight line. Place a sheet of plastic over it and when softened, bend or twist the pliant plastic into a new shape without any molds or forms. When it cools, your part retains its new shape. Simple, fast, and elegant.

The fact that hand-forming thermoplastics requires no expensive molds or tooling featured prominently in the history of toy design. During World War II, plastic materials were used to make bubbles and windshields for fighter planes and bombers, so their availability was limited. At the same time, Elliot and Ruth Handler were just starting their new business making picture frames and desk accessories. Being resourceful, given the limited materials and lack of a shop, Elliot heated scraps and leftover bits of acrylic in his kitchen oven, and, by cleverly twisting and bending the strips, created this swell doll furniture. Out of this modest beginning grew Mattel Toys, one of the largest toy companies ever!

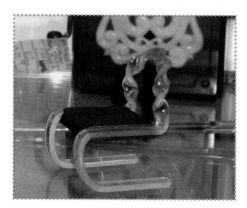

PHOTO COURTESY DEREK GABLE

Make It!

I used a pre-wired heating element made by BriskHeat. It comes with directions for mounting the element in a plywood channel, lined with heatproof fiberglass and foil. I'd recommend using adhesive-backed foil tape instead: it adheres to the plywood and also traps the fiberglass's fraying edges.

1. Cut the wood into three long pieces, one 42″ × 6″ for the base and two for the top, each 36″ × 2⅞″.

2. Center the two smaller pieces on top of the third to make a long trough ¾″ wide and screw into place.

3. Attach two layers of strips of the foil tape along the channel. Lay the fiberglass mat on top and staple the fiberglass to the plywood. Be sure to lay the fiberglass down and into the trough—that's the space for the heater strip. Staple the cloth to the wood.

4. Use aluminum tape all around the edges of the cloth to fasten it to the wood. This will help keep the cloth in position and trap any fibers on the cut edges of the cloth.

5. Lay the heater strip into the trough and attach a wood screw about an inch from each end of the base. Tie the strip's strings together over the screws to hold the strip straight and just snug. Route the connecting wires from each end and join them to assemble the power plug. Attach one more wood screw to the foil tape on the base connected to a long wire. Connect the other end to ground on a 110V outlet.

Your strip heater is ready to try!

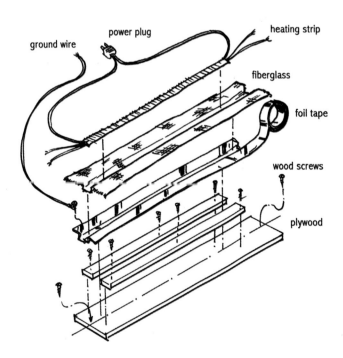

ground wire · power plug · heating strip · fiberglass · foil tape · wood screws · plywood

Easy Easel

Materials

12″ × 24″ ⅛″-thick ABS plastic (acrylonitrile butadiene styrene is a blend of plastics that gives it a tough, non-brittle property. It's used in plastic toys and product housings. Find this at your local plastics supply company. They may have an odds-and-ends bin full of all kinds of leftover trimmed pieces—look there for inexpensive materials! I used black ABS with one textured side and one smooth side.)

Tools

Strip heater
Heat gun
Straightedge
China marker/grease pencil
Hobby knife
Acrylic scribing knife
Edge scraper
Needle-nose pliers
Piece of 1″-thick wood board for bending for bending and pressing
Work gloves

I saw this fun artifact in the window of an antique shop: a mid-twentieth-century easel for reading the morning paper at the breakfast table. It was cleverly made out of a single piece of bent aluminum with silk-screen graphics. On the front was a minimalist mustachioed figure with the words "BREAKFAST WITH FATHER" in a campy retro-1890s typeface. The back simply said: "WIFE'S SIDE." Perfect for today's hipster ironic breakfast. But when I went back to buy it, it was gone!

So here's a version you can make easily at home from one piece of ABS, heated and bent into shape.

Make It!

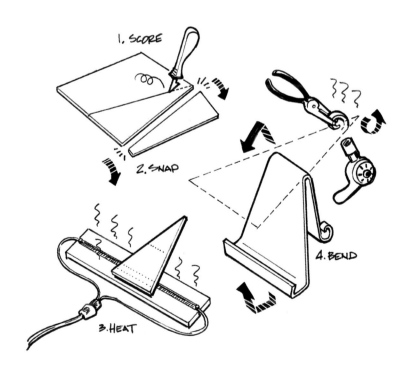

1. SCORE
2. SNAP
3. HEAT
4. BEND

Measure an equilateral triangle with a height of 20″ and a base of 8″ on your piece of ABS. Mark your score lines with a china marker. Dimensions aren't critical so you can modify to use what material you have or to make various sized easels.

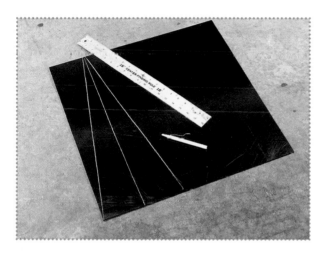

Use the scribing knife and straightedge to make a deep groove along the sides of the triangle. It may take a few passes to do it.

Bend the plastic over the edge of a table groove-side up to cleanly snap out the triangle shape. Use the scraping tool to smooth all the edges of the triangle.

Plug in the strip heater and let it warm up. Use basic and usual safety precautions: don't leave the plugged-in strip unattended. Have a fire extinguisher at the ready. Always heat plastics in a well-ventilated area.

While the heater is warming up, mark your bending lines with the china marker. Draw one line 1″ from the base, another one 2″ from the base, and a third 11″ from the base. If you're using textured material and want to see the texture on the face of the easel, make the 11″ line on the back (non-textured) side and the other lines on the textured side.

Before you bend your easel, try making practice bends on some scrap pieces of plastic. You'll get a good feel for how long to heat

a given thickness of plastic (test it by giving it a test wiggle—is it softened yet?). Make your bends with the heated side of the plastic on the outside of the bend. If left too long or too close to the heater, the plastic could warp, scorch, or bubble. Next time, keep the plastic a little further from the heating strip and make your bend sooner.

You'll make the bends in order from longest to shortest. Place the triangle over the heater, textured-side-up, with the 1″ bending line directly over the heater. Wait until the plastic is softened. Test it carefully by giving it a little test bend, but be careful—don't touch the really hot part of the plastic. Wear your work gloves!

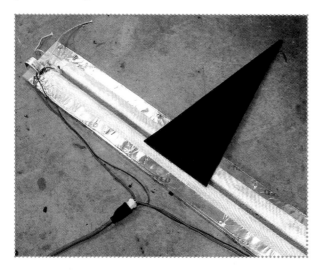

When softened, remove the triangle and place it textured-side-down with the 1″ line over the edge of your work surface. While

holding the triangle in place, press the base down to make a 90-degree bend. Use the wood board to press down against the edge of the table to make a neat, even bend. Hold until cool.

To make the second bend, place the triangle back over the heater with the textured-side up and with the 2″ line directly over the strip. Let soften.

When ready to bend, put the triangle on your worktable, textured-side-up. Place the edge of the 1″ wood board between the already-bent edge and the 2″ bending line. Press the board down firmly as you lift the triangle to make the second 90-degree bend. That makes the "trough" of the easel. Hold the triangle up against the face of the bending board until cool.

and bend it back at a 45-degree angle. Hold until cool.

To make the curled tip, use the heat gun to soften the tip of the triangle. Then give it a twist with a needle-nose pliers and hold until cool. Done!

For the middle bend, place the triangle back on the heating strip, this time textured-side-down with the 11″ line over the strip. When it's softened, remove the plastic

Like the original collectible that inspired it, this handy easel holds your morning paper or tablet at the breakfast table or you could use it on the counter as a cookbook stand. It also makes a nice *Make:* magazine holder on your workbench.

➜Not a Toy—But Still Fun!⟵

This sample project makes a part for a sports car and so is not strictly a "toy" for a kid (more of a toy for an adult!), but I include it here to show the versatility of the strip heater and the process for you to follow to make your own projects from scratch: creating a quick cardboard mock-up, checking sizes and clearances, planning bends and joints, fabrication techniques, and so on. Once you see the simplicity and the application of the plastic bending and joining techniques, you'll think of your own ways to use them. Maybe you'll create a bean bag toss game, or a squirt gun target, or something more practical.

When my son Reed replaced the battery in his Lotus with a new larger one, he discovered that the original, smaller battery cover wouldn't fit. We used the strip heater to fabricate a bigger cover and had fun in the process.

To start, I made a rough cardboard model. It's always a good idea to do a quick mock-up of a new design. You can check sizes and clearances as you go. Much easier to make changes by cutting or adding cardboard. You'll also be able to envision the details of construction, like which corners are bent as opposed to glued together, and so on.

I designed a simple two-part cover and laid out the parts on a sheet of 0.090" textured black ABS using a white china marker. I then drilled ½" holes to make radiused inside corners for stress relief. I scored the cuts with a sharp utility knife and bent them to snap apart. To de-burr the sharp edges, I scraped them with the knife.

Be sure to plan your sequence of bends so that you always have a flat surface to heat. Don't inadvertently bend yourself into a corner.

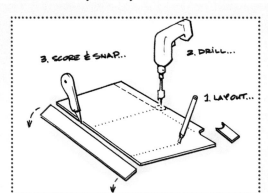

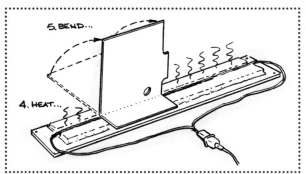

I bonded the parts together with MEK solvent, available at your local plastic supply store. You can paint it on using a paintbrush or get an applicator bottle with a needle tube tip.

Keep the joint you're gluing horizontal so the water-thin MEK flows evenly and doesn't puddle. Always work with plenty of ventilation!

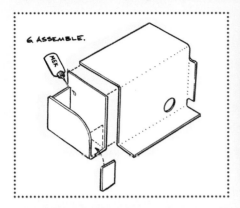

A U-shaped bend in the back of the cover creates a storage compartment, together with a few sidepieces. Self-adhesive Velcro spots hold the removable cover in place inside the Lotus Elise's tiny trunk.

Now . . . what cool, original project will *you* make with your strip heater?

Display Stands

It's always nice to be able to display your collectible toys and art in an attractive way. Art galleries and museums often create custom display stands out of clear acrylic. You can make your own nice-looking stands using the strip heater and a propane torch. With the torch you can "flame polish" the acrylic to give a perfectly smooth and clean edge. Pro shops use high-temperature hydrogen-oxygen torches for professional results but you can get by with your household propane torch.

Here a simple design for displaying PEZ dispensers. You can adapt the design by making it wider to hold more or changing the shape and proportions to display other toys or things. The display can stand on a table or shelf, or can be attached to a wall with two screws.

1. Layout

Lay out the design on piece of ¼" thick acrylic. The protective film makes it easy to measure and mark the locations of the drilled holes. The large holes will accept the PEZ feet. The small holes are for mounting screws. The outside corners will be radiused.

2. Cut and Drill

You can scribe and snap thinner pieces of acrylic, but thicker acrylic over ⅛" will need to be cut. Use a special saw blade with fine carbide teeth to minimize chipping. You can cut on a band saw, saber saw, or circular saw. Clamp your work and keep an even cutting speed and feed. Look out for marred edges where the acrylic is melting due to friction!

Drilling acrylic can also be tricky due to its brittle nature. If you will be doing a lot of plastic drilling consider getting some special drills at your plastic supply shop. The filed-down edges of the drills prevent them from catching in the plastic as the hole is finished. Be careful when drilling large holes: keep the speeds low and the feeds easy and intermittent to reduce frictional heat from galling up the holes. Always place a scrap piece of wood underneath the sheet to be drilled to minimize chipping.

Cut and sand the radiused outside corners. I used a disc sander.

Leave the film on while cutting and drilling to protect the face of the acrylic. When all these operations are done, carefully peel the film off both sides before flame polishing and bending. Carefully sand any rough edges or use the scraper to smooth out the edges before final polishing.

If you're lucky enough to be near a Maker space, you could try having your acrylic laser-cut for you. In that case you'd have nearly unlimited options for cut-out shapes, intricate designs, and fine detail, way beyond anything that could be cut manually!

3. Flame Polish

It will take a while to build up your skills with the torch, so be sure to practice on edges of scraps and leftovers until you are confident enough to polish your work piece.

Wear safety goggles and gloves. Hold the work piece in a clamp or vise, or you can hold it in your hand. Position the edge so it faces up and you can see it clearly. Light the propane torch and adjust so it has a fine pencil tip and adjust for a medium-blue flame. Stroke the flame across the only the edge of the acrylic. Move smoothly and continuously—don't linger! You want to melt the plastic just enough to make a glossy, smooth finish on the edge, but no more! The plastic can very quickly bubble or burn. With practice you'll find the right speed and distance to produce an even, smooth finish. Of course, the entire time you must also be *very* careful with the flame and keep it away from any materials that could ignite!

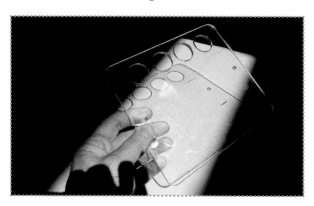

4. Bend

For thin sheets of acrylic, you can heat from either side, but for thicker pieces you must heat the outside of the bend. It needs to be softer since it's bending around a larger radius than the tighter inside of a corner. Also, thicker acrylic needs a little more time on the strip heater to heat through.

For this project, the exact location of the bends isn't critical, but you do want them to be parallel and each 90 degrees. You can check each bend by placing the shelf on a flat, smooth surface and checking the bend with a small square. Before the bend cools, you can make minor adjustments. I like to work on a big mirror: the reflection helps me see when things are truly vertical and square.

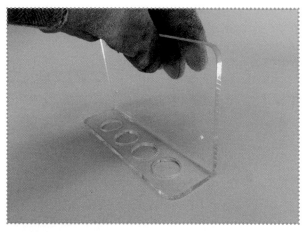

Here are some other display stands I made. Note the eye-shaped hole to hold the carving. You can cut or grind any shape you need to fit your object d'art!

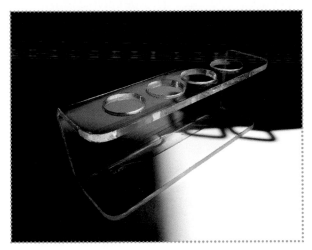

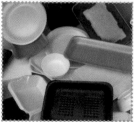

Five-Minute Foam Factory

What keeps your coffee warm, but also rides the Pacific surf? What's used to cushion delicate audio gear in shipping, but makes an annoying, squeaky sound? What's banned in over 100 cities, but you can find it just about everywhere? It's expanded polystyrene (EPS) foam. It's a great insulator (for that coffee cup and wall insulation), lightweight and stiff, and impervious to water (great for surfboards)—but it's also impractical to recycle and can be an unsightly part of the waste stream. Our landfills and waterways are filling up with used coffee cups, grocery store meat trays, and tossed-out take-out packaging. With this easy-to-make foam cutter you can reuse leftover EPS foam to create treasures from trash!

Because EPS is a thermoplastic foam it can be cut with a hot wire like a knife through butter. You can buy expensive commercially made hot-wire cutters, but here's how to build a quick, super-simple DIY design for next to nothing, and how to get great results with some clever accessories and foolproof techniques.

Materials and Tools

Scrounge your workplace for scrap materials to build the cutter. None of the dimensions are critical so feel free to adapt the sizes shown to use what you have.

2 × 4 (or any dimensional lumber), two pieces 18" long

18" × 18" pegboard (any thickness, tempered or not)

21"-long, ¼"-diameter aluminum rod

2 test leads with alligator clips

Train transformer (one with a variable DC control is ideal)

Nichrome wire (0.010" diameter with a resistance of 7 ohms per foot. Get it at a hobby store or online.)

Bolt and 4 square nuts (any size)

Wood handsaw

Hacksaw

Drill and bits

Screwdriver

Tape measure

EPS foam material

(Optional/not shown: multimeter with ohm setting)

Sure, you can buy it at a craft store, but why not get creative and recycle? Once you start looking, you'll find lots of EPS foam in everyday items you'd otherwise throw away:

- Grocery store meat trays (in cool colors like black, blue, red, and yellow!)
- Picnic plates
- Coffee cups
- Fast-food containers
- Packing materials
- Styrofoam coolers
- Leftover chunks of house insulation (Dow "blue-board" Styrofoam™ and OwensCorning InsulPink® are both *extruded* EPS [not made from little beads] and give excellent results!)

For the cutting wire, this design uses a fine wire made of nichrome (nickel-chromium). It's held vertically on a table and kept taut by a bent aluminum arm. A model train transformer converts AC power into a controllable and safe 12 volts of DC power to warm up the wire. This hands-free design lets you guide the pieces of foam into the stationary wire and slide the foam around to make effortless cuts.

Safety Warning

When heated, EPS can produce benzene, and when burned, gives off other noxious fumes. Be careful: *Always* use your hot wire cutter in a very well-ventilated area!

Use the train transformer's speed control to set the temperature of the wire: just warm enough to cut the foam, but no warmer. *Never* cut with a red-hot smoking wire!

Foam made by Woodland Scenics is specially manufactured so that it doesn't give off any harmful fumes when cut with a hot wire. Get it where model train supplies are sold or at the Woodland Scenics website (woodlandscenics.com).

Be careful when cutting foam with a hot wire. You don't want to touch it—it's over 200°F! Be extra careful and wear gloves!

1. Make the Base

Measure and cut the pegboard to make a tabletop. Mine is 18″ × 18″.

Measure and cut two pieces of 2 × 4 to make leg rails.

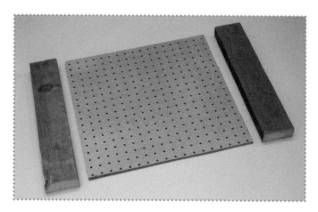

Use screws to attach the top to the rails. Place a ¼″ drill bit through one of the pegboard holes over the middle of a 2 × 4 and drill all the way through the rail. This will be the hole that the rod fits into.

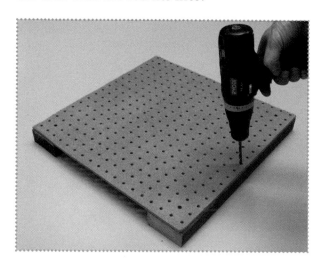

2. Bend and Insert the Rod

Cut the aluminum rod to 21″ in length. Mark off 12″ and bend the rod about 90 degrees.

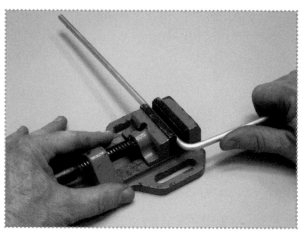

Insert the short leg of the bent rod in the drilled hole. Mark the position of the closest pegboard hole beneath a point an inch or so from the tip of the rod. Then mark the top surface of the rod directly above the marked hole.

Use the hacksaw to make a shallow notch across the top of the rod where it's marked.

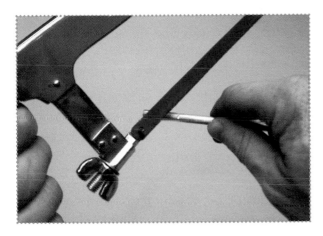

Insert the rod and drill a small ¹⁄₁₆" pilot hole through the rail and into and through the rod.

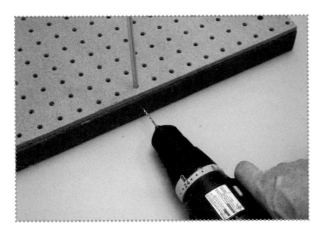

Drive a nail into the hole and through the rod. This prevents the rod from swiveling in its hole as you cut the foam in different directions.

3. Add the Wire

Thread two nuts onto the bolt. Wrap, then tie, the nichrome wire around the bolt, and then add the remaining two nuts. Tighten the nuts to pinch the wire.

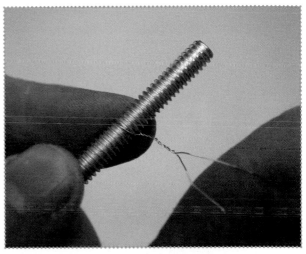

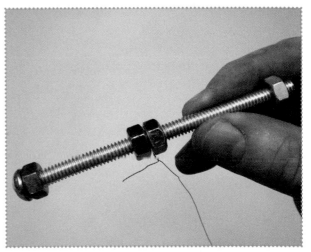

Align the nuts so that they all lay flat. Put on a drop of superglue to lock it tight.

Thread the nichrome wire through the marked hole and pull it up vertically.

Press the tip of the rod down slightly and hold it there. At the same time, make a loop in the wire so that the tip of the loop just reaches the deflected rod. Hold that loop and tie it off in a simple overhand knot.

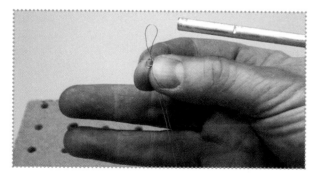

Press the arm down and slip the loop over the rod and into the notch. Let go. The gentle spring force of the rod should make the wire taut. If it's too loose, shorten the wire by tying another knot. Trim any stray ends.

4. Check the Circuit

Let's check the circuit. You might remember this formula from your high school physics class: Voltage equals resistance times current.

$$V = I * R$$

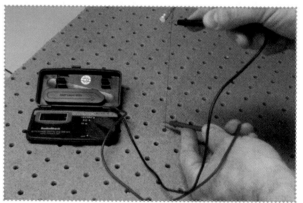

You can measure the resistance of the wire with a voltmeter: set it to *ohms* and measure the wire by placing a probe at each end point. My wire measures 7 ohms. My transformer puts out 12 volts DC. Plugging that into the formula gives you:

$$12 = (I) * 7$$

or

$$12 / 7 = I$$
$$I = 1.71 \text{ amps}$$

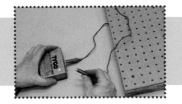

The current needed, therefore, is just under two amps. My train transformer is rated at two amps, so that's good, at least for short time periods. The resistance of the wire will change at various temperatures, so the current drawn will vary. Many train transformers have a built-in thermal breaker—if they get too warm, they'll shut off. If that happens, unplug the transformer and let it cool down. It should work again later.

5. Power Up!

Use the alligator clip lead to attach the train transformer. Unplug the transformer. Connect one clip from one side of the regulated DC contacts to the bolt underneath the table. Connect the other lead to the remaining DC volt contact on the transformer. Make sure the variable control is on its lowest setting, then plug in the transformer.

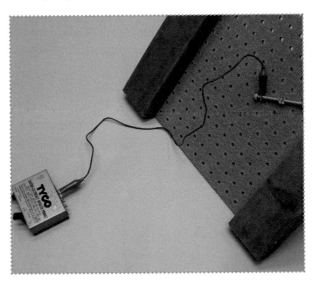

Ready to test it? Lastly, connect the unconnected alligator clip to the base of the rod. You've created a circuit that sends current through the wire. Adjust the transformer's control so that the wire gets warm—not glowing red-hot. No heat? Check your connections and make sure the clips aren't touching at the transformer.

Test your cutter with a scrap of foam. Place the foam on the table surface and gently slide it into the hot wire—it should cut easily with just the slightest pressure. Adjust the voltage if needed. Don't press too hard, though, or you'll pull the wire into an arc and your cuts will be curved instead of straight.

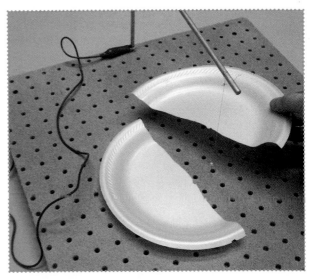

Use that connection to the rod as your on/off switch. You can see when it's connected and the hot wire is "on."

Cutting Techniques

You can cut free-form shapes easily—just keep the foam moving at a smooth, constant speed. You'll notice the slightest wiggle in your movements will result in a wavy or ridged part!

Try the following simple fixtures and guides for easy cutting and perfect parts.

1. Straight Cuts

Make a guide from a piece of 1 × 2 and some ¼" dowels. Drill two holes on 1" centers and slip in pieces of dowel. Plug the dowel pins into the pegboard and you've got an adjustable fence for smooth, straight cuts.

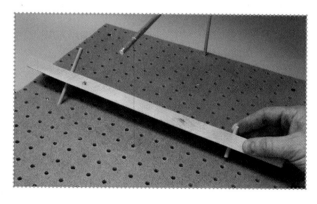

Position the fence close to the wire to cut thin strips.

Position the fence farther away to make wide cuts and to square up blocks.

For size adjustments of less than 1", remove one of the dowel pins and use the remaining pin as a pivot. Swivel the fence around to adjust the angle until the distance between the wire and the fence is just right, then place a second long dowel pin behind the fence at the closest pegboard hole. Fast and easy!

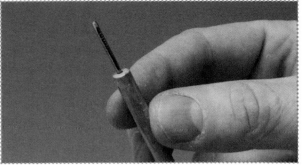

2. Circles

Make a circle cutter guide out of a wood dowel and a nail. Cut a dowel to a length so that when it's inserted into a pegboard hole, it's flush with the tabletop. Drill a ¹⁄₁₆″ pilot hole and insert a small nail—head first. Tap in with a hammer. You have a dowel with a pointy pin sticking out—be careful!

Place the dowel in any pegboard hole. The distance from the wire to the pin will be the radius of your circular cut. Energize the wire. Slide a piece of foam into the wire and impale the foam onto the pin. As you spin the foam around on the pin, you'll cut a perfect circle. Turn off the wire and remove your part.

3. Angled Cuts

You're not limited to 90-degree cuts. Build an elevated and angled wedge table and you can make beveled cuts!

4. Conic Sections

Combine a raised and angled stand with a circle-cutting action and you can cut cones! Make a wedge block as before with an angled top and dowels on the bottom to fit the cutter table. Then add an upside-down nail as a pin on the top angled surface.

Spin the foam around the angled axis and you'll cut a circle with an angled side—a cone.

5. Surfaces of Revolution

You can actually cut compound curved surfaces with a straight wire! A cone cutter with an offset axis will create a *hyperboloid*, the same shape as a power station cooling tower. Use an angled pin, just as with cutting a cone, but position it so that its axis of the pin and the cutting wire are skew lines: non-intersecting lines in different planes. Spin the foam carefully around the pin. The angled axis will cut a twisted shape. Of course you'll have to cut or break apart the surrounding ring of foam to free your part. Try various angles and distances between the pin and the wire to make gently or tightly curved shapes.

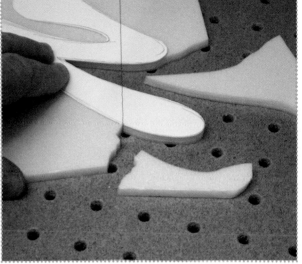

You'll get a smooth and accurate cut. If you happen to stray and cut wide of the edge, just go back and cut again. Use the template over and over for identical perfect parts, again and again!

6. Templates

For perfect results with complex shapes, make a cardboard template to use as sort of a "slide guide." Cut your shape out of thin cardboard and pin it or tape it to the foam using double-sided tape.

Slide the hot wire along the edge of the cardboard as you cut the foam.

Photo Stand-ups

This simple project makes a cut-out "paper doll" from a favorite photo. Print out a picture of a figure on a sheet of label stock. Carefully cut out the figure from the background.

Peel off the back and stick it to a piece of EPS foam (a black-colored meat tray looks cool!).

Then, use the paper figure cut-out as a slide guide as you cut the foam with the hot wire. The wire won't cut through the paper.

Make a notch at the bottom and add a small stand—makes a fun stand-up figure!

Foam Plate Flier

"Upcycle" your used foam picnic plate into a flying falcon! See the appendix for a clip-and-use pattern.

1. Trim the Pattern

Glue the pattern onto thick paper or card stock and carefully cut out the two shapes, one for the wings and body and one for the tail fin. Be sure to cut out the thin slots in the pattern for the tail and for the ailerons.

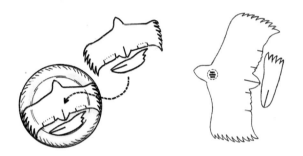

2. Cut the Foam

Pin or tape the pattern in place on a foam plate using double-sided tape. Fire up your foam cutter and cut away. If you make a mistake and cut a little wide of the pattern, just go back and trim again. Be sure to cut the slots, too.

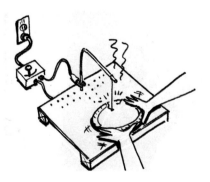

3. Assemble

To assemble, slide the tail into the slot as shown. Add a penny with double-sided tape for a nose weight. Bend the ailerons up a bit and adjust the tail to be perpendicular. The curved shape of the plate makes a cool dihedral on the wing tips.

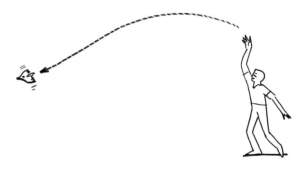

4. Fly!

To fly, toss gently overhand from a high point. The falcon should glide down. Adjust the ailerons, tail, and position of the weight if needed.

Identical Snowflakes

Since it takes very little force to cut with a hot wire, you can cut a stack of foam pieces to make a bunch of identical parts, all at once. Try this: make a snowflake design by folding a 6-inch diameter chipboard circle and cut your design. Leave the side edges uncut. That way there'll be no closed "holes" in your snowflake design.

Unfold and pin to a stack of Styrofoam picnic plates.

Carefully cut along the guide (go slowly!). You've made a stack of identical snowflakes! Make some small slits at the outside edge of the snowflakes.

You can connect them slot-to-slot to build a snowflake sculpture.

Halloween Skull Slices

You can also make multiples of varied thicknesses of any shape by using the ripping fence.

1. Cut the Shape

First I cut a skull shape. If you have a design that has an unconnected hole, like the eyeholes in this skull, you can cut them by first poking a small pilot hole through the foam with a pencil.

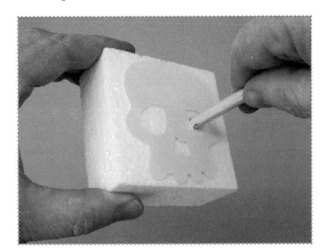

Unhook the nichrome wire from the rod, thread the wire through the hole, and reconnect the loop over the notch.

Turn on the hot wire and cut the hole!

2. Cut the Slices

When finished cutting the holes, disconnect the wire, remove the shape, and reconnect the wire. Place the ripping fence in position, turn the shape on edge, and then cut slices to create skull-shaped Halloween decorations.

Double- and Triple-Cut 3D Shapes

You can also create 3D shapes by making a sequence of two cuts at 90 degrees to each other. This example makes a cute teddy bear figure.

1. Draw

First, draw matching front and side views of the bear's 3D shape onto some thin cardboard. You can also download the pattern online and print it out. These will serve as cutting templates.

Use the guide fence to square up a block of foam, then carefully double-side tape the front-view and side-view templates on the foam. Be sure to line up the patterns.

2. Cut the Front View

Cut along the pattern with the hot wire—but be careful to leave all the cut pieces in place as you go.

I've temporarily removed the shape to show the result of the first cut.

3. Cut the Side View

Put the bear profile shape back in the block and tape on the side view. Be careful to register the side view's position with the previous cut and keep the cut-off portions together in place. Turn the block 90 degrees and make the side-view cut.

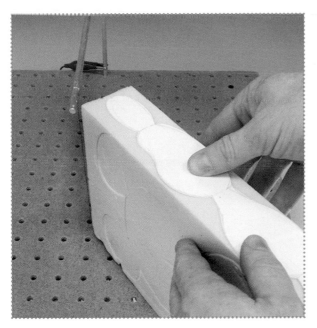

This exploded view shows the bear shape surrounded by its cut-off parts.

Carefully remove the trimmed-away sections to see your 3D shape.

Here are more ideas for things to make with cut foam: signs, picture frames, floating pool-party decorations, gliders and boats, rocket nose cones and fins, boomerangs . . . what shapes will *you* make with your hot-wire foam cutter?

Chapter 5

Casting and Molding

The poor incandescent light bulb, the very symbol of having a bright idea, is endangered. It's being phased out all around the world: first in Brazil and Venezuela in 2005, then in the European Union in 2009, and now in the United States. Granted, efficient CFL and LED lights produce more light with far less energy but they're just not as much fun. The light bulb in the E-Z-Make Oven creates enough heat to shrink plastic, harden polymer clay, fuse enameling powders, and make wiggly creations by "cooking" plastisol. So, have your own aha moment and build this light bulb–powered craft oven—while you can!

E-Z-Make Oven

The E-Z-Make Oven design is super simple: a light bulb in a can. Holes in the bottom serve as cool air intake and are also used for mounting the insulated feet and bulb socket. A grid of holes in the lid allows the hot air to rise up around the molds. An inverted pan serves as a cover to hold in the heat. The oven reaches about 300°F with a 75W spotlight. That's just enough to do several different kinds of heat-activated crafts: harden polymer clay, cure plastisol, shrink plastic, and fuse low-temperature enamels.

Maker Tip

WARNING: The side of the oven can get hot, so be careful and use the cool wire handle to lift or move the oven.

Materials

Empty, clean 1-gallon-size paint can with lid and handle (buy one from the hardware store)

75W reflector incandescent spotlight

Bulb socket, lamp cord (4′), power plug, switch

¼″ inside diameter grommet, bolts and nuts (to fit your socket), synthetic wine corks (for feet), small screws

Tools

Compass
Center punch
Drill
Step hole drill bit
De-burr tool
Screwdriver

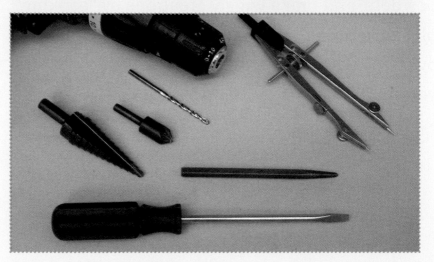

1. Lay Out the Holes

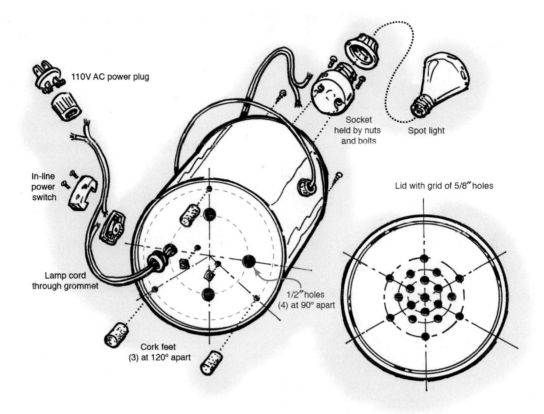

110V AC power plug

In-line power switch

Lamp cord through grommet

Cork feet (3) at 120° apart

Socket held by nuts and bolts

Spot light

Lid with grid of 5/8″ holes

1/2″ holes (4) at 90° apart

For the holes in the lid, find the center, and then use the compass to scribe a circle with a ⅜″ radius. Use the compass set at that radius to mark off six holes and center punch each of them. Repeat with 1″ and 1½″ radius circles. Exact spacing isn't critical, but you'll want a regular distribution of holes for even heat flow in the center of the lid.

Lay out the holes in the bottom of the can following the diagram.

Set the compass to the radius of the can, then use the compass to mark off distances around the rim to locate three equidistant holes for the feet. At each location, mark a spot ½″ inside the rim and center punch each spot.

Place the lamp socket in the center and mark the locations of the mounting holes. Center punch them. (If your socket has a single screw mount in the center, mark and punch a center hole instead.)

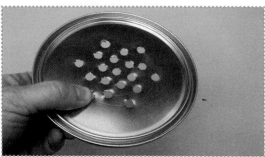

Lastly, set the compass for half the can's radius, scribe a circle, and mark off four equidistant ½" holes and center punch them.

2. Drill the Holes

In the bottom of the can, drill three ⅛" holes for the feet, four ½" holes for venting, and holes to fit the small bolts for mounting your lamp socket. Drill the grid of holes in the lid. Debur the holes as needed.

Maker Tip

A step drill works well for drilling thin metals, where a twist drill tends to grab and pull up the metal. It's easiest to use the step hole drill bit with a drill press: control the diameter of the hole by the depth of the feed.

3. Mount the Hardware

Cut two synthetic corks in half and drill a ⅛" pilot hole in three of the halves. Attach the three half cork feet using small screws pushed out from inside the can.

Insert the grommet in one of the four ½" holes and thread the lamp cord through it. Tie a knot in the cord 6" from the end, then split and strip the wires. Connect the wires to the socket.

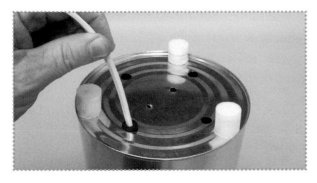

Bolt the lamp socket to the inside bottom of the can using the bolts and nuts.

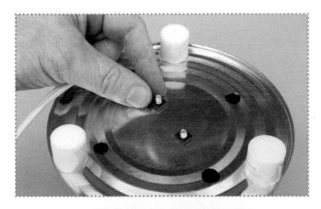

Install the in-line switch on the cord, 12″ from the can, following the directions from your switch.

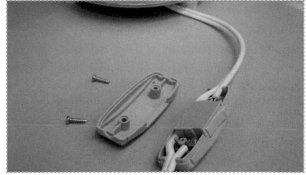

Finally, strip the other end and attach the wires to the AC plug. For a polarized two-prong plug, the narrow (hot) blade goes to the center (hot) contact on the socket. For a three-prong grounded plug, the bare or green wire goes to the green ground contact on the socket, if any.

Screw in the bulb and test.

4. Make the Cover

Invert an aluminum-foil loaf pan and attach a synthetic cork to the outside. Drill a ⅛″ pilot hole in the cork and fasten to the pan with a screw.

Custom-shaped Metal Casting Mold

Unlike thermoplastics that can be melted and remolded, molding with thermoset materials is like making a hard-boiled egg: once cooked, its form is permanently set. Plastisol starts out as a creamy liquid that cures into a soft, pliant plastic. Baby Boomers might remember Mattel's *Thingmaker* toy, which came with die-cast zinc molds. (See Chapter 4.) You can make your own custom-shaped molds out of easy-to-form aluminum.

The form is a hard "positive" pattern over which you'll stretch and shape the aluminum sheet to make a "negative" mold. Make your form by sculpting soft polymer clay, then fire the clay until firm using the E-Z-Make Oven.

Materials

Small aluminum loaf pan, or thick aluminum tooling foil (available at hobby stores)

Sculpey polymer clay
Masonite hardboard

Tools

Foil-burnishing tools (make your own from hardwood dowels, with various sizes of rounded tips)

Sculpting tools

1. Sculpt the Form

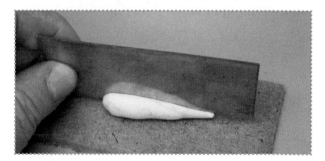

Cut a piece of hardboard on which to make your form. Knead a small quantity of polymer clay until pliant and then sculpt your shape. Keep in mind that you'll be stretching and deforming the aluminum over your form, so keep your shape rather shallow (the aluminum sheet will only stretch so far) and allow plenty of draft (angled sides without undercuts). You'll be amazed at the small details that the foil will pick up. This example shows what happens when you burnish the foil over a penny!

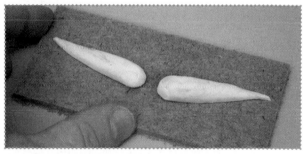

Here I'm making a mold for casting moustaches. Roll the clay into a teardrop, then place it on the hardboard. Cut in half lengthwise with a thin piece of metal or a knife, shape the curl, and add two bits to make nose pinchers. Lastly, sculpt in some hair detail. Take some time to make small features and textures in your polyclay form.

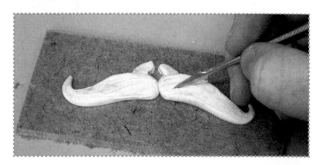

When your polyclay form is finished, bake it in an oven set to 275°F for about 20 minutes or until hard. Probe gently with a toothpick: when the clay springs back, it's done.

You can also fire your polyclay using the E-Z-Make Oven you just built. It gets hot enough, but it will take more time than using a big kitchen oven. Place the lid on top of the form to keep in the heat.

When it's done, set your firm form aside to cool.

2. Tool the Mold

The material used for the mold must have three properties: a low specific heat (to efficiently conduct the light bulb's limited heat to the plastisol), to be soft and easily formed, and compatibility with the plastisol's chemistry. Steel is too hard, copper is not chemically compatible, but aluminum has all three attributes. Get some 0.005″ thickness tooling aluminum from an art supply store. It's specially made to be soft and is easily worked with wooden tools. You can also use pieces cut from a disposable aluminum pan. It's a little stiffer and harder to form, but will work fine.

Cut a piece of foil somewhat larger than your form.

Lay the sheet over the form and press the foil around the form with your fingers. A few wrinkles and folds are inevitable, but you'll work them smooth later.

Make some tools from wooden dowels of various diameters. Leave one end flat and make radiused tips on the other end in various sizes from gently rounded to pencil-point sharp.

Use the wooden tools very carefully to press the foil closer to the form. Don't try to stretch the foil all at once: be careful not to tear it or poke a hole. Instead, work slowly from the sides or top toward the center of a valley or crease. Start with the roundest tipped tools first. Do a little at a time. Slowly approximate the shape, stretching the foil as you go. Easy does it! This may take some practice until you get the hang of it.

Once you have defined the basic shape, go back with a finer-pointed wooden dowel and work the small details. Firm-but-gentle pressure gives the best result. Often, as you get to the deepest part of the form, the foil will be very thin and tear easily. No worries: your mold may still be quite usable, since you don't fill the mold up to the very top. Keep working!

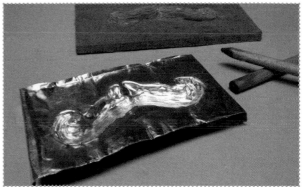

Use the flat end or the side of a dowel to smooth out any wrinkles and make a flat surface surrounding your mold shape. Finally, fold down two sides to make legs so the inverted mold stands level.

Cast Plastisol Parts

A good source for bulk plastisol and pigments is Industrial Arts Supply Company (IASCO-TESCO.com). Get formulation #16 for nicely soft and rubbery molded parts. You can also get pre-colored ready-to-use "goop" in squeeze bottles from Patti-Goop on Amazon or eBay. I've even used old bottles of Mattel's original *Thingmaker* Goop—it still worked after 50 years. Now, that's swell!

Place the mold on the oven and add the cover to preheat. Fill a squeeze bottle with plastisol and pigment and mix thoroughly. Fill the mold by squeezing plastisol in slowly and letting it fill from bottom to top to avoid trapping any bubbles.

You can even "paint" your part by using different colors: let one color set before adding another or pour together to get a swirl effect. Place the lid over the mold to help trap the heat.

Thin shapes cure quickly; deep shapes take more time. Probe with a toothpick: when the plastisol is firm, it's done. Carefully remove with tongs or tweezers (it's hot!) and let cool.

When cool, use a toothpick to carefully pick at an edge, then lift and peel your part out of the mold. If you're careful you can cast many parts from the same mold.

→Casting a Mold←

There's no limit to the customized or personalized shapes you can create. In this example my daughter Laura Knetzger molded this cute beetle character based on her own 'zine comic, *Bug Boys*. But instead of burnishing foil over a male form, for this shape it was burnished into a female form, which was made first by casting resin. Use this technique when starting from a soft original (like clay or a found object) to make a mold that is hard enough to burnish the foil.

First, a positive is sculpted from soft clay. Plenty of detail can be included in the clay.

Next, the form is sprayed with mold release and liquid resin is cast around the clay in a mold tray.

When the resin sets, the clay and tray can be removed, leaving a female mold.

The soft metal foil can be formed as before, but now the metal is stretched down into a female mold.

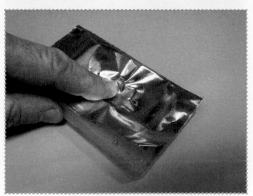

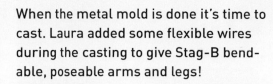

When the metal mold is done it's time to cast. Laura added some flexible wires during the casting to give Stag-B bendable, poseable arms and legs!

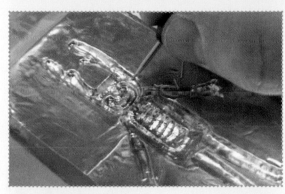

Faux Enamel Trinkets

Materials

Brass sheet
Low-temperature
 enameling powders
Cardstock for stencils
Bead chain necklace or
 jewelry pinback

Tools

Saw
Sandpaper
Center punch
Drill
Scissors or hobby knife
E-Z-Make Oven (or
 kitchen oven)

Use the E-Z-Make Oven to create fun faux jewelry and trinkets! This project features colored powders that melt and fuse at low temperatures to give a bright, shiny finish. They look like glass enamel or cloisonné but you can bake it with just a light bulb. Find the fusing powders in craft stores or online. Embossing powders sold for stamping will work, too.

1. Make a Metal Shape

Cut a small disc or any other shape out of brass and debur the edges with sandpaper. Make a mark with a center punch and drill with a ⅛" drill bit to make a hole near the rim. Clean the brass with vinegar and water to remove any finger oil so that the powders will adhere uniformly. Keep it clean: only handle the trinket by the edges as you go.

2. Add Powders and Fuse

Sprinkle on some low-temperature powder and build up an even layer about ⅟₃₂" thick. Then carefully place the trinket on the warmed E-Z-Make Oven. After a few minutes when the powder melts and gets shiny, remove it and let it cool. You can add more powders in different colors to create a design.

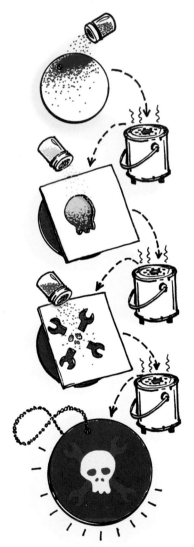

I made this Jolly Hacker medallion with hand-cut stencils of thin cardstock. You could also make super-intricate, laser-cut stencils. When all your colors have been added, return the trinket to the oven for a 20-minute final bake for maximum hardness (about 300°F in a kitchen oven).

Add a chain to finish your medallion, or glue on a pin back to make a brooch or badge.

→ Salute to Makers! ←

Douglas Stith was inspired to take his uke-building project to the next level. He used this process and low-temperature enameling powders to decorate the cool cake-pan uke he made. Here Anthony and Gianna enjoy the beautiful uke!

<small>Photos courtesy Douglas Stith</small>

Custom TiddlyShrinks

Here's something else you can do with your E-Z-Make Oven: shrink plastic to make customized tiddlywinks.

Tiddlywinks is played with a set of small, color-coded discs. Use your *squidger* (a large plastic disc or shape) to press down and scrape across the top surface of a *wink* (a small disc). When you snap its edge, the wink flips up into the air. With a little practice you can aim your wink to land into the target cup.

Create your own custom squidgers and winks using shrink film. (The shrinking film is sold under the Shrinky Dink® name or as a similar generic brand at hobby stores.)

1. Make Your Image.

Draw and color your own design with permanent markers. It's easy to copy your favorite design: place the clear film on top of the original and trace. Or create your design on the computer and use the special shrink film made for ink-jet printing. Just remember, your design will shrink in size by about one half!

2. Cut It Out.

Then cut the shapes out with scissors or a hobby knife. You can also use paper punches to make shaped holes. Be sure to completely cut out your final shape: once the plastic shrinks it will be too thick and tough to cut!

3. Shrink!

Shrink the plastic using your light bulb–powered E-Z-Make Oven. (You can also bake in your kitchen oven at 325° or use a heat gun.) As the thin polystyrene film shrinks, its thickness swells to $\frac{1}{16}$". Heat slowly and evenly to prevent curling. When it's cool, you'll have tough, strong plastic winks and squidgers of your own design in high detail!

4. Play!

Set up a tiddlywink course with a shallow target cup on a piece of felt or tablecloth. Play a round of "tiddlywink golf:" how many tiddles will it take to get your wink into the cup? What is the farthest distance away you can be and still get your wink into the cup on one shot? Look online for many variations and strategies in tiddlywinks game play. Try for a *squop* (landing your wink on another player's wink, freezing it from play) but don't *scrunge* (bouncing out of the target cup—oops!).

The sky's the limit for your own custom tiddlywink graphics and themes: sports, movies, comic books—or mash them up. When you make your own you're not limited to the official licensed versions: Star Wars Angry Bird Basketball tiddlywinks, anyone?

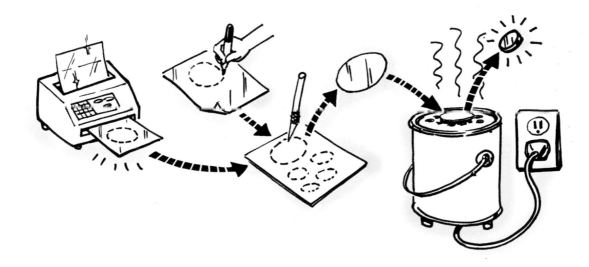

Guitar Amp Bulletin Board

This project uses two different kinds of casting: flexible room temperature vulcanizing (RTV) mold making and hard resin cast parts.

Sometimes the leftovers from one project can be an inspiration for something else. I had an odd piece of vintage speaker cloth that was too small to use on an amp but too cool to throw away . . . hmmm . . . how about using it to make a mini "guitar amp" bulletin board?

To give it a cool look like the front of a guitar amp, I made a framed tray with quarter round wood on a piece of plywood. For the look of Tolex guitar amp covering, I covered the frame with textured black vinyl upholstery material and attached it to the tray with contact cement.

For the bulletin board, I cut some masonite to make a thin backer board to fit inside the frame. I used staples around the back to fasten the speaker cloth around the board.

I lifted some Fender amplifier script art for a real retro look and printed and mounted it on another strip of thin board. This served as the "control panel" storage for the "knob" pins.

Here's the casting part: the things that look exactly like guitar amp knobs are really pushpins! I used RTV silicone to make a mold

around an actual knob, then cast acrylic to make perfect copies.

RTV is a kind of silicone rubber. When mixed with a catalyst, it hardens from a thick syrup into a firm but flexible rubber. I placed an actual knob face up in a small mold tray and poured the RTV on top. I removed the knob after the rubber had set. You can see the mold has captured every tiny detail from the numbers to the ridges and recesses.

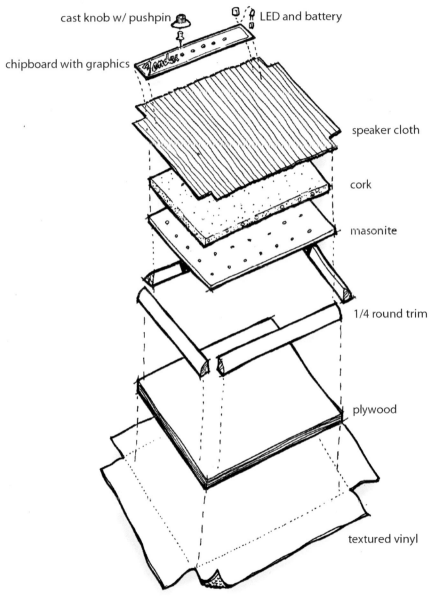

cast knob w/ pushpin

LED and battery

chipboard with graphics

speaker cloth

cork

masonite

1/4 round trim

plywood

textured vinyl

To cast the knobs, I mixed up some polyester resin with MEK catalyst and carefully poured it into the RTV mold. I tapped the mold to release any bubbles. Then I pushed a pin through a small piece of cardboard and placed it upside down over the mold, suspending the pin in the uncured resin.

Maker Tip

WARNING: Always work with plenty of ventilation when casting polyester resin. The fumes are dangerous!

Later I unmolded the cured knob and sprayed it with black paint. I wiped it with white paint to fill in the engraved number markings. I painted the center recess of the knob with silver paint. This made a perfect copy of the original knob!

I drilled some small holes for the knob pins in the control panel. Looks just like a real amp!

I finished it off with a working pilot light by wiring a micro switch on the back of a button battery holder and wired it with a super-bright red LED. I used a cap from a tube of lip balm as a bezel. When you press the pilot light pin into the board, it glows red. Rock on!

You can do a lot with RTV rubber molds—they're so flexible! I've used them to make molds from all kinds of found objects and cast different materials from resin to chocolate to Sculpey. The inherent flexible rubbery property of the silicone means it will flex away from any undercuts in the original shape and in the part you cast when you unmold. It's also very durable and with care you should be able to get many parts out of an RTV silicone mold. For more details on using RTV silicone and polyester resin, consult the instructions that come with each.

Edible Optics

Gelatin is used in lots of things like food and candy, and pharmaceutical and paintball capsules, but it also has interesting optical properties. Fun fact: Before the advent of plastics, gelatin was used to make colorful filters for theatrical lighting and photography, and these filters are still called *gels* today. A fun way to play with gelatin optics is to cast lenses.

Find a cup or bowl with a smooth curved inside to use as a mold. If you have any chemistry equipment, a concave watch glass makes a perfect lens mold. You also can make a convex lens mold by stretching plastic wrap over the mouth of a glass and then sucking out the air before quickly sealing the wrap. The stretched film will make a wrinkle-free concave shape. Add a rubber band around the rim to keep it sealed airtight.

Mix up some gelatin and leave to set in the mold. Mix the gelatin in a very concentrated form, about 1Tbsp of powdered gelatin in ⅛ cup of hot water. Stir until dissolved.

Pour the gelatin in the mold and let set in the fridge.

When set, remove the wrap to use the lens. Unflavored gelatin makes clear lenses, but you can use flavored gelatin to play with colored filters, too.

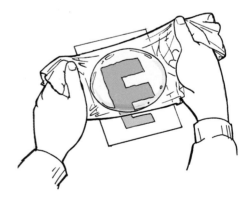

How powerful is your lens? That depends on the shape and the material:

$$1/\text{focal length} = (n-1)\,(1/\text{lens radius})$$

With a refractive index of around n = 1.38, gelatin is less refractive than glass (n = 1.51) and only slightly more than water (n = 1.33). I'll leave it to you to combine two gelatin lenses to create a "Jell-O-scope." When you're done looking, enjoy your edible optics. Looks good enough to eat!

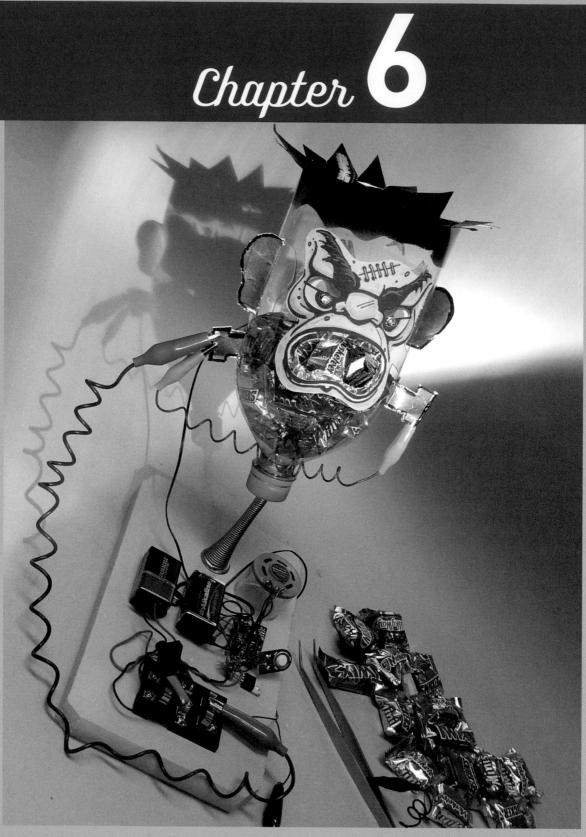

Chapter 6

Holiday Projects

Holidays are a fun opportunity for making your own toys, games, and crafts. You can use them as seasonal decorations, play party games, or even give them as personalized gifts. It's the thought that counts, so think about making some of these!

Mad Monster Candy Snatch Game

Here's a classic toy reimagined for Halloween candy-giving and party fun. It's the *Mad Monster Candy Snatch* game, which combines the nerve-racking dexterity of the game *Operation* (BZZZZZT!) with a monster head–shaped candy dispenser. Put the fun in *fun-sized candy* and make your little goblins earn their treats with this tricky game!

The see-through green monster head is filled with fun-sized candies. Do you dare to snatch a snack? Use the forceps to carefully reach inside its mouth. If you can maneuver out a candy, you've won a treat! But be careful—if you touch the side, you lose. The monster wakes up with crackling, shocking sound effects and announces "YOU MAKE MONSTER MAD! YOU LOSE!!" as his angry eyes flash red. No treat for you!

Materials

Large green-colored soda bottle (I used a ginger ale bottle)

Doorstop spring (get the kind as shown that has a tapering large-to-small conical shape for just the right amount of bendiness)

Aluminum tape (not silver-colored duct tape—REAL metal tape!)

6 alligator clip jumper wires

Long metal tweezers

DPDT knife switch (RadioShack part 275-1537)

Sound recording module (RadioShack part 276-1323)

Wire

2 super bright red LEDs

TIP31 NPN power transistor

220-ohm resistor

9V battery clip with leads

Small piece of perf board

Double-sided foam mounting tape

Wood for base, approximately 6″ x 10″ x 1″ thick (anything will work: particle board, plywood, or solid wood)

Black and yellow paint

Screws from your hardware jar

Masking tape

Adhesive-backed blank label sheet

Clip-and-use for holes and face label (See the appendix for clip-and-use pages or download files at makerfunbook.com to print out.)

Halloween candies (fun-size or mini candy bars, or any small, wrapped candy you can pick up with tweezers)

Tools

³⁄₁₆″ punch (to fit your LEDs; you could also use a drill)

Soldering iron and solder

Scissors

Screwdrivers

Drill and bits

Hobby knife

Black marker

It's simple to make, and you can customize it to make it as easy or difficult to play as you like. You can even personalize it with your own voice, choice sayings, and sound effects.

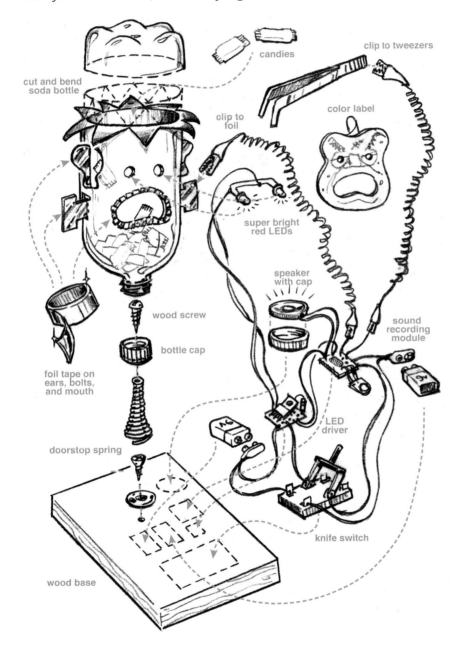

candies

clip to tweezers

cut and bend soda bottle

clip to foil

color label

super bright red LEDs

speaker with cap

wood screw

sound recording module

bottle cap

foil tape on ears, bolts, and mouth

LED driver

doorstop spring

knife switch

wood base

1. Make the Monster's Head

Empty the soda bottle and save the cap. Rinse it out and remove the label. Cut off the bottom of the bottle to make a big opening. Jaggedly cut the edges to form the monster's spiky hair; don't be too neat, he's a mess! Mask off the remaining part of the bottle (again, don't be too precise). Paint the jagged tips of the bottle with black spray paint. When dry, bend each of the triangular points outward to make the pointy "hair."

Find the candy monster pattern page in the appendix. Cut out on the thick black dashed lines. Fold over on the thin fold lines. Wrap the pattern around the bottle, and using a black permanent marker, trace the cut lines onto the bottle. Use a hobby knife to carefully poke a starting slot. Using sharp scissors, cut along the lines. Punch out the eyeholes with the ³⁄₁₆″ punch to fit the LEDs. Fold back the ears and neck bolts at 90 degrees so they stick out.

Cut a piece of aluminum tape ½″ x 6″ and cut slits ¼″ all along one long side. Next cut more slits on the other long side, alternating the cuts so you don't snip the strip all the way through. Then stick the tape to the inside edge of the mouth hole: place the uncut center of the tape all along the edge, folding over onto the outside and inside of the bottle. It should create a foil-lined edge all along the mouth opening. Cut more pieces of aluminum tape and stick to both sides of the ear flaps and neck bolts. To finish the head, go to makerfunbook.com and download the file for the face label, and print it out on an adhesive label sheet. Cut it out along the dotted line and be sure to cut out the eyes and mouth. Carefully center it over where the eyes and mouth should go and adhere it to the outside of the bottle.

2. Make the Base

Cut the wood panel to size. I painted mine yellow.

Remove the rubber tip from the doorstopper and find a round-head wood screw that will be a good, tight fit inside the end of the spring. Drill a hole on the center of the bottle cap just big enough for the screw. Thread the wood screw from the inside of the cap and tighten the screw inside the spring.

Drill a pilot hole and fasten the doorstop's mounting base into the center of the wood panel. Twist the spring until it fits tightly. Twist the upside-down bottle onto the cap. Test the spring by filling the monster head with a couple handfuls of candy and give it a push. The candy-filled head should deflect and wobble but not bend over.

3. Make the Circuit

First, test the sound circuit; add a 9V battery. Press and hold the REC button and speak clearly into the microphone. Let go of the button to stop recording. You can record up to 20 seconds of sound. If you like, go to makerfunbook.com to the sound prerecorded just for this game. It has crackling sound effects, zapping lab sounds, and the monster saying: "YOU MAKE MONSTER MAD! YOU LOSE!!" Hold your phone's speaker right up to the circuit's microphone and press and hold the REC button as before to record this sound into your circuit. (Or, if you prefer, you can instead record your own sounds and voice.)

You'll need to slightly modify the sound-playing circuit so that the game's tweezers and foil sensors trigger the PLAY button contacts. Here's the modified circuit for the game.

Scan this QR code to hear the monster.

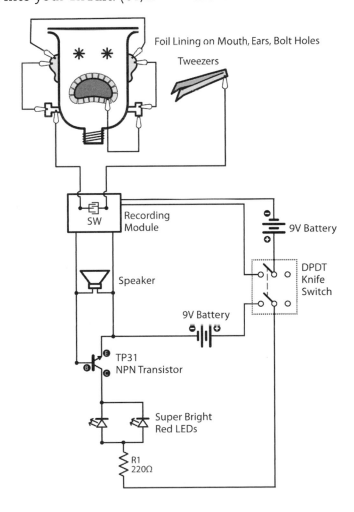

Foil Lining on Mouth, Ears, Bolt Holes

Tweezers

SW

Recording Module

9V Battery

Speaker

DPDT Knife Switch

9V Battery

TP31 NPN Transistor

Super Bright Red LEDs

R1 220Ω

Here's the unmodified circuit board.

Carefully remove the PLAY button from the circuit board by prying up the metal tabs on the back of the board. Remove the metal retaining ring and the gray elastomeric button.

Insert two small wires through the holes on either side of the switch pad (SW) and very carefully solder one wire to each of the two traces. Don't short out the traces! When you touch the ends of the two wires together, it should trigger the sound player. Test it!

Solder up the rest of the circuit for flashing the LEDs: Add the power transistor to the perf board, and after noting the E, C, and B legs, wire up the connections to the speaker, then solder the connections to the dropping resistor and the two LEDs. Wire the LEDs up in parallel about 2½" inches apart on a 12" long lead that will reach into the monster's head. Wire up the 9V clip, add the second battery, and test the circuit. The LEDs should flash brightly as the sound plays. If not, check polarity on the LEDs, transistor, and battery.

Use a small cylinder or plastic cap from a milk bottle to make a resonant chamber for the little speaker. It won't take much to improve the naked speaker's tinny sound. Superglue the speaker to the cap.

You don't really need the knife switch but it adds a cool "mad-scientist's lab" look to the monster. Use some wood screws to mount the switch to the board. If you want, you can wire the switch to turn the power on and off. Just wire each half of DPDT in series with each 9V battery. See the wiring diagram for details.

Use the double-sided adhesive foam tape to mount all the components to the base.

4. Assemble the Game

Screw the head back into the spring/cap base. Twist the bottle a little if needed to make the mouth face forward.

Thread the wired-up LEDs in through a neck-bolt hole and insert the LEDs into the eyeholes from the inside. They should fit snugly, but if not, tack them in place with a bit of hot glue or superglue. The wire should be loose and not restrict the bottle from bobbling.

Now connect everything together using alligator clip jumpers. For an even more mad-scientist look, wrap the wires tightly around a pencil first to give them a "coiled cord" look.

Connect one of the wires from the PLAY sound trigger to the tweezers. Connect the other PLAY sound trigger wire to foil on a neck bolt; then use more jumpers to connect the foil mouth, foil ears, and the other neck bolt all together.

Final test: Close the switch and touch the tweezers to the foil on the edge of the mouth; the monster should talk and flash his eyes! Test the other contact points on the neck bolts and ears with the tweezers, too.

Load up the monster head with some fun-size candies and you're ready to play!

How to Play

Easy game Reach into the monster's mouth with the tweezers and try to get a candy without "waking the monster" (touching the sides). BZZZZZT!—your turn is over; pass the tweezers to the next player. If you've succeeded, eat your candy or add it to your trick-or-treat bag! You can make the game easier to win by simply cutting larger holes for the ears and neck bolts.

Simple strategy Add a die or spinner labeled *Mouth*, *Ear*, and *Neck*. On your turn, spin the spinner and try to snatch a candy from the opening indicated. If you're successful, you can try again—but if you miss, you lose all your candies—put them back into the monster's head and let the next player go! Will you risk it or play it safe?

Name that candy The player to your right names which specific candy you must try to get. You may have to do a little careful digging with the tweezers to win!

Another Salute to Makers!

It's great to see what people come up with for these projects as printed in *Make:* magazine. The *Mad Monster Candy Snatch* game inspired some readers to make it their own!

I'm not sure if there is a merit badge for candy snatching, but Techno Teams at Girl Scouts Heart of the Hudson tried their luck at Girl Fest Celebration 2014 in Mahopac, New York.

GIRL SCOUTS HEART OF THE HUDSON, TECHNO TEAMS

The Schiffmans in California built their version as a family. Jacob Schiffman had the great idea of substituting his Snap Circuits toy for the sound-making electronics. Check out his clever hack!

His mom Julie says, "We had so much fun playing the game. Thank you so much for inspiring our family to make our own games—our kids truly love developing their own game ideas!" Note how they used the Snap Circuits for the base. For year-round, sugar-free fun, they swapped the candies for LEGO bricks. Here's brother Eli taking his turn.

Scan this QR code to see a demo video of the Mad Monster Candy Snatch game in action!

PHOTOS BY RICHARD SCHIFFMAN

Magic Moving "Ouija Be Mine" Card

Want to send your Valentine a really special message next February 14th? Here's an animated pop-up card you can make for Valentine's Day. When you slowly open the card, a hand sweeps across and moves the heart-shaped planchette over the Ouija board, spelling out the message: "HAPPY VALENTINE'S DAY!" It's both mysterious *and* romantic, perfect for any secret admirer to make and send.

Materials and Tools

White cardstock, 8½" x 11"
2 brass bindery brads
Hobby knife
Optional: ⅛" punch and mallet

1. Print and Cut

It's also easy to make: just go to makerfunbook.com and download the two graphic files or use the images in the appendix. Print them on card stock or laminate a paper version onto thin cardboard.

Cut the hand and long tab out on the solid lines. Don't forget to cut out the circular hole in the planchette! Use a small ⅛" diameter punch to make the two small holes in the hand

and one hole in the tab. If you don't have a punch, carefully cut the small holes with sharp hobby knife. Check to make sure the tabs on your brads fit inside the holes. If your brads are bigger, enlarge the holes.

Score and slightly bend the tab on the dotted line.

Trim off the bottom part (unused) and gently score and fold the body of the card on the dotted line. Cut the slot and the small hole in the card with a sharp hobby knife.

2. Assemble

To assemble, use a brass brad to fasten the tab to the hand, then thread the tab through the slot in the card. Finally, fasten the hand to the card with a second brass brad. (The brads even look like two little brass buttons!)

Check the action by opening and closing the card and make any fine adjustments to the brads or re-crease the folds for smooth and easy movement. Hope this card will help you "make" it a Happy Valentine's Day!

I can't guarantee you'll get the same result if you make the "Ouija Be Mine" Valentine, but Seattle writer Steve Wacker made one and gave it to his girlfriend Betsy—and now they're married! They sent me this sweet selfie.

Scan the QR code and see the Ouija Be Mine card in action!

Custom Cookie Cutters

Make your own custom-shaped cookie cutters—two ways! One way is to cut and bend metal strips. The other way uses the Kitchen Floor Vacuum Former to make plastic cookie cutters. You can make metal ones pretty quickly, but if you want to make multiple cutters of the same design, you could mold several on the same vacuum form.

Draw your cookie shape outline to use as a same-size pattern. Keep it simple; small details are okay but a bold silhouette will work best. Narrow shapes or delicate features will break apart and be frustrating to cut, bake, and frost.

Here's an example of how to create a successful cookie cutter design. My daughter Laura wanted to try making some cookie cutters based on her comic book characters, Rhino B and Stag B, aka, The Bug Boys. You can see them at the top of the page.

As drawn, the Bug Boys comic legs and horns might be too thin for cookies. Let's simplify the design and modify the proportions a bit.

Materials and Tools

0.015″ thick aluminum sheet
Pop rivets and tool
Needle-nose pliers
Drill and ⅛″ drill bit
Hobby knife
Sandpaper
Ruler

Metal Cookie Cutters

For metal cookie cutters use 0.015″ aluminum sheet, available at hardware stores. Measure 1-inch wide strips, then score firmly with a hobby knife. Bend the score against the edge of a table, then bend back and forth until it splits. Repeat to make as many strips as needed. Sand the edges to remove any sharp burrs.

Use needle-nose pliers to hold the strip and carefully make bends to match your pattern. Use a pencil or dowel inside to form tight corners. The soft aluminum bends easily and holds its shape. Keep your bends perpendicular to the strip!

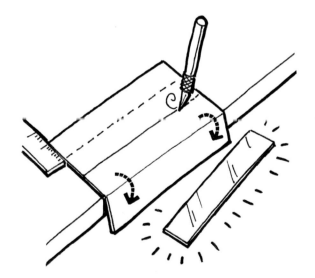

To make longer strips or join the ends, drill matching ⅛″ holes. Use pop rivets with backing washers to fasten strips together. Done!

Let's try them. Laura's making ginger-bread cookies. She rolls out the dough and presses down to cut.

How'd they come out? Pretty good—making the arms legs and horns thicker makes it easy to cut and bake.

Plastic Cookie Cutters

To make plastic cookie cutters, first cut your shape out of wood, then vacuum-form plastic over it. (See Chapter 4, "Thermoforming Plastics," on how to build and use your own Kitchen Floor Vacuum Former!)

1. Make the Form

Transfer your design to the foam or wood. The cutters don't have to be too tall but allow a little extra height, since you will later cut off the top of the form to make a cutting edge—1″ high should be plenty. Give the form some draft where you can to make it easier to get the form out later.

Materials and Tools

Foam or wood for the form
0.030″ thick styrene sheet for vacuum forming
Hand or hobby saw
Hobby knife
Sandpaper
File
Optional: Surface gauge

2. Mold the Plastic

Vacuum form some 0.030″ plastic (thicker if you can get it!) over your form. It's okay if there are some webs or places where the plastic doesn't closely stretch over the form. You only care about the very top of the shape—that's the part that will cut the cookie dough.

3. Trim

Cut the formed plastic all around the shape, leaving a generous rim. The rim will give the plastic part strength. You can cut with a saw or just score and snap off as you go around the shape.

Use the surface gauge to scribe a mark all around the top of the form, just below the top face of the shape. If you don't have a surface gauge, use a small pencil. Turn the shape over with the face flat on the table. With the pencil also flat on the table, slide the pencil along so it makes a mark all around the side of the shape at the same height.

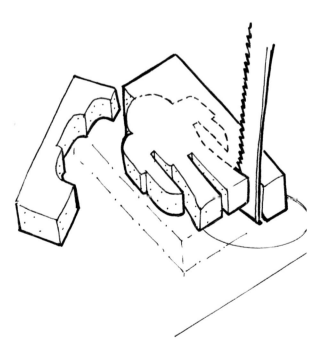

Use the hobby saw to cut the face off the top of the part where marked. Keep your cut even and flat. To finish it off, place a sheet of sandpaper face up on a tabletop and place the cookie cutter on top, flange side up. Slide the cutter back and forth, around and around to sand the top cut flat.

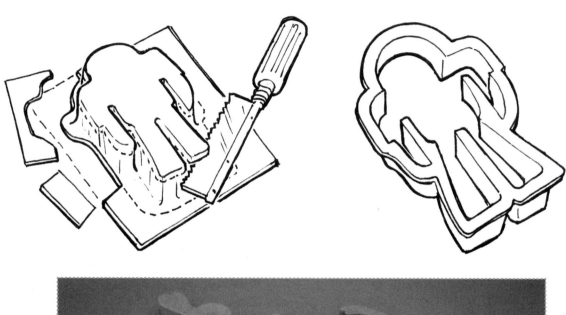

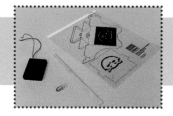

Magic Cat Projector

Make this fun mini projector toy for Halloween. It throws a shadow image of a black cat—which then magically disappears, leaving only his spooky grin!

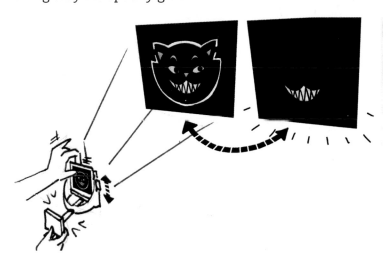

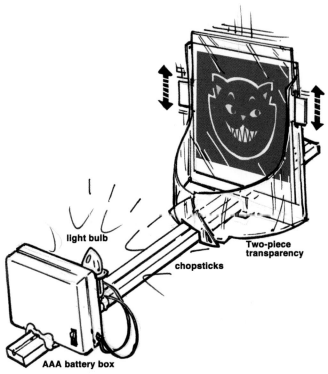

light bulb

Two-piece transparency

chopsticks

AAA battery box

Magic Cat Projector (continued)

1. Print and Cut

Go to makerfunbook.com to download the image file. Print it out full size on a single sheet of transparency film. Or use the clip-and-use in the appendix and make a copy on clear transparency film.

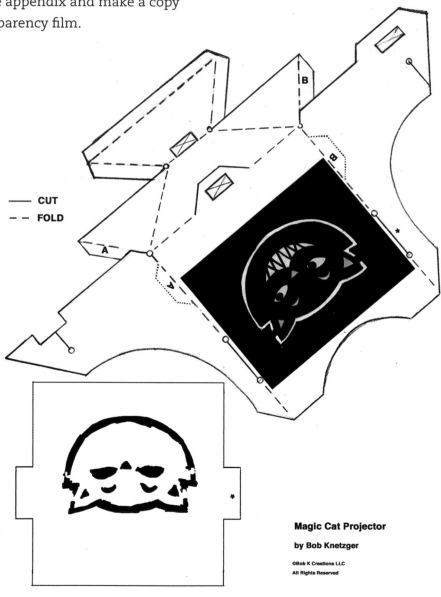

—— CUT

- - - FOLD

Magic Cat Projector

by Bob Knetzger

Use a hobby knife to cut out the two shapes. Carefully cut the slits as shown and use a small-sized (⅛″ hole) paper punch to make the little circles to provide stress relief at the inside corners.

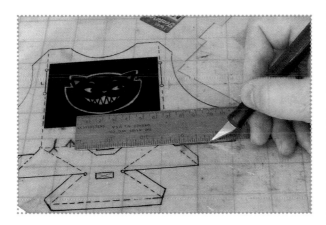

2. Assemble

Score on the dashed fold lines by tracing over them with a ballpoint pen and straightedge.

Then fold as shown. The sides fold forward and interlock.

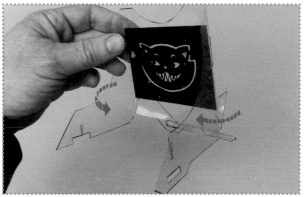

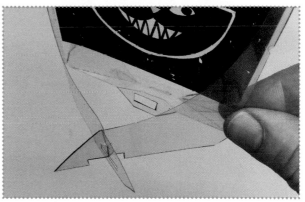

Fold the bottom backward and bend the two wings up. Bend the A and B tabs in. Tape the tabs to the center at A and B.

Thread an unbroken take-out chopstick through the square holes.

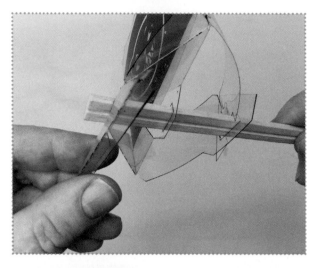

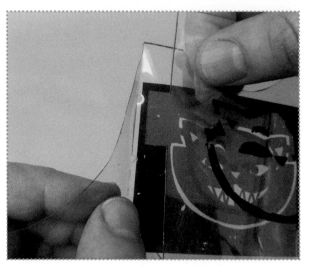

Solder the two leads from a 4 AAA battery pack to the contacts of a PR13 flashlight bulb. The battery box I used had a built-in on/off switch. If yours doesn't, just take the batteries out to turn it off.

Match the asterisk on the slider, and center and insert the ears of the slider into the slits in the center. It should freely slide up and down in the slits.

Hot-glue the battery box to other end of the chopsticks.

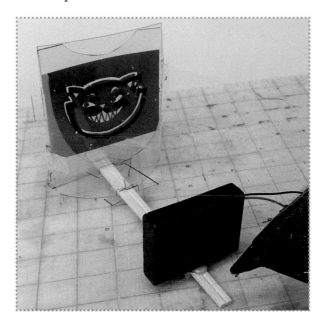

Dim the lights and slide the film up and down to make the cat disappear and reappear!

Hot-glue the light bulb to the chopstick and to the backside of the battery box.

Scan the QR code and see the Magic Cat Projector in action!

"*Gnome* Math Required" Holiday Hats

Materials and Tools

String
Felt-tip pen
Push pin
Straightedge
Cardboard working mat
Felt 24″ square piece for each hat
Scissors
Sewing machine

Once upon a Christmas time I had to come up with some stocking stuffers. Whatever it was, it had to be simple, easy, and fast, since I wanted to make a bunch of them for friends and family. Our backyard garden gnome was the inspiration: I'd make felt gnome hats for everyone. And all I needed to make the pattern was a little geometry and a piece of string!

I knew the basic shape was a cone and could be "rolled" from a circular piece of felt, but what dimensions should I use?

I made a quick orthographic sketch with an auxiliary view to find out. I envisioned a cone, not too squat or too pointy, that was about 10½″ high. The vertical dimension in the frontal view is not really the radius (it's foreshortened), but by making an auxiliary view that is perpendicular to the projection lines, the edge view of the side of the cone becomes the true-length radius of the true-size flat circle. Now I knew the radius dimension: 12″.

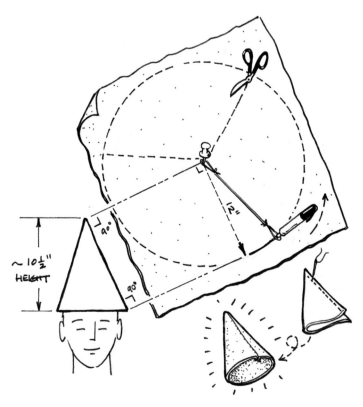

1. Mark the Pattern

To lay out the circles on the felt, tie loops on each end of a piece of string so that the distance between the end of the loops is 12″. Place the felt on the cardboard and put a pin in the center. Put one string loop over the pin and put the other loop around the marker tip. Keeping the string taut, draw a circle.

An average 7½″ hat size has about a 23″ circumference ($2\pi r = C$). So, each felt circle will yield three hats. Use the string loops to divide the circle into thirds. Put the pin on a point right on the circle, then use the string and marker to measure, and tick-mark a point along the circumference. Place the pin in the new mark and make another mark along the circumference. Do it again and again to make six marks all around the circle. The marks won't be perfectly spaced, but that's okay.

Use the straightedge to draw a radius line from the center to one of the marks on the circle. Draw two more radii to make three "fat pie pieces."

2. Cut and Sew

Cut out the three pie-shaped pieces of felt. Fold a felt piece over, matching the straight sides,

and sew a seam to join the edges. One quick run through the sewing machine and it's done!

Turn the felt inside out and presto: a holiday gnome hat! You can whip out a stack of hats in no time for everybody on your holiday list!

With the leftover felt pieces, I made mini-sized gnome hats for friends, family, and pets, too.

PHOTOS BY DAVID CORINA

PHOTO BY PAT FULLAM

Chapter 7

Working with Metal

*H*ere are three more "meta projects"—projects you build first, then use them to make *other* projects. These all use metals and metal-working processes in a fun way

The Desktop Foundry project makes a cool miniature foundry that casts tiny metal parts. It uses Field's metal, a special, safe, low-temperature casting alloy. You could try substituting other low temperature alloys, but be careful. Don't use an alloy that contains lead or cadmium.

I dressed up the Desktop Foundry with fancy brass parts I made using a process called photo etching. Even if you don't etch any parts yourself, it's interesting to read about the process and see examples of fantastically detailed parts made with it.

The last project in the chapter, the Yakitori Grill, is especially satisfying: it makes dinner. Everyone will love Ninja Chicken Sticks!

Desktop Foundry

Build this miniature working foundry and cast real metal parts safely, right on your desk. Make custom jewelry, tiny trinkets, die-cast-style game tokens—then remelt them and recast, again and again.

What makes it all possible is a special *eutectic* alloy—one with the lowest melting point—called Field's metal, which melts at an amazingly low 144 °F (about the temperature of hot coffee). Unlike other metals with a low melting temperature, this alloy of bismuth, indium, and tin contains no lead or cadmium and is safe and nontoxic.

Maker Tip

CAUTION: Even though Field's metal melts at a very low temperature, it can still burn you if you touch it while it's in its liquid state.

The basic foundry is made from wood and metal along with a few scrounged household parts. If you're up for a bit of a challenge, the second project in this chapter will show you how you can also dress up your Desktop Foundry with some snazzy brass trim and a twirling phoenix turbine.

Scan this QR code with your smartphone to see a demo video of the foundry casting a metal Makey Bot trinket!

Project Overview

You can vary the dimensions and features to suit whatever materials you have on hand. None of the materials or dimensions for this project are critical except for getting your hands on some Field's metal. It's a little pricey but you'll only need a thimbleful or two of the stuff to have fun casting, melting, and recasting.

Fun fact: It's named for its inventor, author and *Make:* contributor Simon Quellen Field!

Use Sugru, the super-easy-to-use silicone rubber material, to make the molds. It comes in handy small pouches perfect for this application. Find Sugru online, as well. Look for more details in the next project, Cast a Trinket.

I pilfered an alcohol lamp from an old chemistry set, but again, these are easy to find online and they are quite inexpensive. For the

Materials

Wood for base, 7½" × 5½" (I used a piece of solid ¾" thick solid oak shelving)

16 gauge soft brass wire, about 2 feet

³⁄₁₆" brass tubing × 5" long

Two ¼" × 1½" 0.030" brass strips

Two ³⁄₁₆" I.D. set screw collars

Small round head brass brads

Small glass bottles

Thimble

Alcohol lamp

1" hardwood dowel (6" long)

¼" hardwood dowel (3" long)

Wood screw and washer

Two corks

Sugru® molding compound

Field's metal (available at www .scitoyscatalog.com)

Self-adhesive rubber feet

Kitchen matches and striker strip

White glue

Wood stain

Non-stick cooking spray

Tools

Hand drill or drill press

Wood saw

1½", 1", ¹⁄₃₂", ¼", ³⁄₃₂", ⅛", ¹⁄₁₆" drill bits

Screwdriver

Ruler

Vice grips

Optional

Ogee curve router bit

Router

Mill and lathe

¾" and ½" end mills

Photo-etching materials

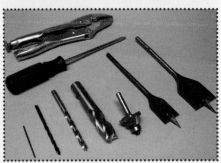

see-through crucible and storage vials, I used contact lens shipping bottles, but any small glass bottles will do. To make it easy to pour, use a bottle with the least "shouldered" neck.

The simple foundry design has a center shaft mounted vertically on a wooden base.

The shaft swivels 90° to move the crucible from the heating lamp over to the mold. The cork handle swivels to tilt the crucible and pour the molten metal into the mold below. There's also storage for molds, metal, and matches. This tiny foundry uses a thimble for a wick snuffer.

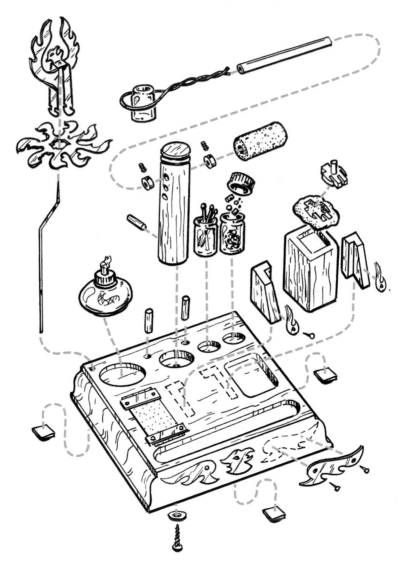

1. Make the Base

Start by cutting the base to size. If you like, use a router to add a decorative flourish to the edges. I used an ogee curve bit for a "desktop pen set" look. I also cut a 45°-beveled front face, but you can leave yours plain.

Mark and drill the holes for the shaft, bottles, and lamps. All holes are drilled ⅛" deep, so use a drill press for best results. Drill a ⅛" through-hole in the center of the recessed shaft hole. I also milled some recessed tray areas for storage of the snuffer, lamp, and extra molds, although you could use a router as well.

Make a mold holder using 2"× 2" or any leftover wood scraps from your shop. I milled a small recess in the top to help hold the Sugru, but it's not required. Cut some L-shaped supports, then cut bevels on the front. When glued in place in the center of the base they'll keep the mold holder in position while casting.

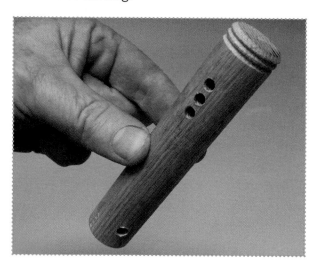

Cut the 1" dowel to 5" length and drill a ⅛" pilot hole in the bottom for the wood

screw. Drill a series of ⁷⁄₃₂" holes for adjusting the height of the crucible (see sketch for details). Twist the dowel 90° and drill a ¼" hole—make it ⅜" deep and ½" from the other end of the dowel. I added some decorative rings using a lathe.

Cut three 1" lengths of ¼" dowel and glue one into the hole near the bottom of the shaft. Place the large dowel in the center

hole in the base and fasten the wood screw and washer through the bottom. Tighten the screw just snug so the vertical shaft can rotate in the hole. Add three stick-on rubber feet to the bottom of the base.

2. Craft the Crucible

Cut a 14″ length of brass wire and wrap the center around the lip of the small glass bottle. Twist the brass wires to make a long handle, then cinch tightly with the vise grips. Thread the twisted brass wire through the brass tube. Put a collar over the tube, then slide the tube through one of the 7/32″ holes in the shaft.

Test the fit with the alcohol lamp in place: turn the shaft and slide the brass tube so that the crucible bottle is directly over the lamp's wick. Hold the collar against the shaft and tighten the set screw on the collar.

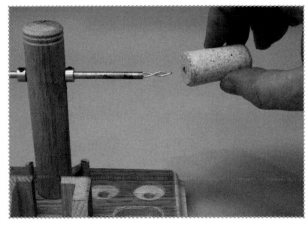

Add the second collar on the backside of the tube and tighten it in place against the other side of the shaft. Drill a ¼″ hole in the center of a cork and twist the cork snugly onto the end of the wires and brass tube to make an insulated handle. The crucible should swivel as you turn the cork.

With the crucible positioned over the lamp, hold one of the ¼" dowel pieces vertically on the base so that it touches the ¼" dowel in the shaft. Mark that position on the base and drill a ¼" hole × ⅜" deep. Glue the second ¼" dowel in the hole. This makes a positive stop to locate the crucible over the lamp.

3. Mount the Mold Holder

Next, swivel the shaft 90° counterclockwise and place the mold holder underneath so that the crucible will pour directly into it. Carefully place the mold supports on either side and mark their position on the base. Also hold a small dowel next to the dowel on the shaft, mark its position and drill a ¼" hole × ⅜" deep. Glue the third ¼" dowel in place—that will make a positive stop for the

shaft in the pouring position. Glue the mold supports to the base, too.

4. Finishing Touches

Cut a 10" length of brass wire and wrap the center around the lip of the thimble. Using a vise grip, twist the brass wires tightly to make a handle as before. Drill a ³⁄₁₆" hole

through a cork and thread it over the end of the twisted wires to make an insulated handle.

Finish the wood parts with a thin coat of dark wood stain to bring out the grain.

Drill some 1/16″ holes in each end of the two small brass strips. Cut a piece of striker material from the side of a box of kitchen matches. Place the striker on the base as shown and put the brass strips on each end. Mark the holes, drill 1/16″ pilot holes, and then gently tap in the small brass brads, holding the striker material to the base with the brass strips.

The basic Desktop Foundry is ready to use. The next project shows you how to easily make a mold and cast parts, but if you want to take your Desktop Foundry to the next level, skip ahead to the project on metal etching. You'll see how to add some extra brass trim and a whirling phoenix turbine to make your Desktop Foundry worthy of your office or den!

Cast a Trinket

Use the Desktop Foundry to cast a metal part. By the way, you can also use this same technique and materials to mold lots of different items. Besides low-temperature alloys you can cast custom-shaped chocolates, resin figures, plaster shapes, novelty ice cubes, and more. Just about anything that you can pour as a liquid and then let set into a solid would work.

1. Make the Mold

Find a small coin, trinket, or object you'd like to mold. I sculpted a tiny *Make:* robot from a piece of styrene and added a dimensional letter M punched out from a plastic label-maker strip. Whether you sculpt your own shape, or use a found object as your pattern, be careful to avoid any major undercuts. Small ones are okay for this project, because the cured Sugru will easily flex when unmolding. Just be extra careful when you remove your pattern, and then gently reshape the uncured Sugru as needed.

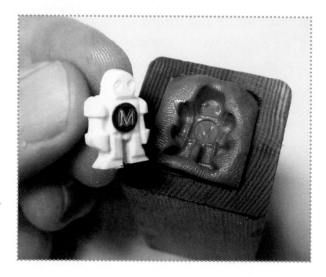

Spray some non-stick spray onto the top of the mold holder (you don't want the Sugru to stick to it). Open up a pack of Sugru and knead it, then press it into the mold base. Spray your object with some non-stick cooking spray as a mold release, wipe away the excess, then carefully press the object into the Sugru. Leave it in place for about 24 hours for the Sugru to firm up.

Carefully remove the object; you'll have a mold with every tiny detail and surface texture reproduced in it. See how Sugru stays flexible and forgiving so that you can easily unmold your cast metal shape?

2. Cast a Part

Clean the oil from the mold and dust it with a tiny bit of talcum powder to help the molten metal flow better. Put your mold in place on the base.

3. Create Jewelry

After you've tried casting a few parts, why not add a few jewelry findings to make something really special—custom jewelry!

After you've poured the molten metal, but before it cools, carefully add an earring post and hold it there until it cools. You made an earring!

Similarly you can add a pin back to create a cool lapel pin.

Place a small quantity of Field's metal in the crucible and swing it into position. Light the lamp. Gently twist the cork handle to swivel the crucible back and forth as the metal melts. When ready, swing the crucible over to the mold and twist the handle to pour the metal into the mold.

Give the mold some gentle taps to help the molten metal flow into any details and to release any bubbles. When cooled, flex the Sugru to unmold the metal part. You've cast a real metal treasure! You can remelt and recast again and again.

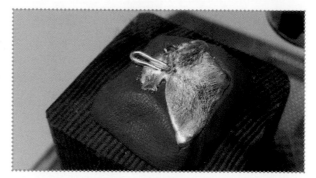

To make a charm, make a small wire loop and add it to a molten casting. When the metal cools, use a few jump rings to add the charm to a necklace or bracelet.

There's another interesting material that you can use to make molds for metal casting, and unlike Sugru, it will even hold up to high temperatures: cuttlebone! It's the internal shell of the cuttlefish and is sold in pet stores as a calcium supplement for birds and reptiles. Its soft structure is perfect for making molds. Just press a hard shape firmly into the cuttlebone. It crunches down to form a perfect impression of the original shape. Jewelers since ancient Egypt have used cuttlebone to cast precious metals.

Photo-Etched Metal Parts

While working on the Desktop Foundry project, I was surprised to learn that my local shops with CO_2 laser cutters wouldn't cut brass. It's too reflective and can damage the lens. What to do? Instead, I thought I'd try to create the intricately detailed brass parts using a different process: photo-etching. Better living . . . through chemistry!

Photo-etching is used to make parts for small-scale model trains. This 1:48 scale model helicopter from Czech Republic comes with a sheet of photo-etched brass parts. Czech out the exquisite detail! Those tiny bolt heads are only 0.010″ in diameter.

British designer Sam Buxton has created some spectacularly clever art pieces for his line of Mikro men. A single flat piece of etched stainless steel is bent and twisted to create a miniature 3D world. My favorite is a bubble-helmeted space man taking his pet Mars rover out for a walk, holding a robotic pooper scooper. There are also half-etched details of textures, lettering, and markings.

Some of the pieces I photo-etched for the Desktop Foundry project were simple, flat, decorative brass trim, but two parts were bent and folded to create a 3D, phoenix-shaped turbine.

If you've ever etched your own circuit boards you may already have most of the required materials. If not, consider using the excellent Pro-Etch kit from Micro-Mark. It has everything you'll need in one box: all materials and chemicals, and an excellent instruction manual.

The process (shown on the next page) is similar to etching a printed circuit board, except there's no board. In short: Print your image on clear film and place it over the photosensitized brass. Expose it to a light source and place it in developer. The areas covered by the black image wash away, leaving an acid-resistant pattern. Place the brass in ferric chloride solution to etch away the unprotected areas, leaving your etched part. Here are the details.

→The Micro-Mark Pro-Etch Kit←

This great kit has absolutely everything you'll need to photo-etch at home, right down to gloves and safety goggles. The detailed instruction manual covers every aspect of the process clearly and is full of great tips and examples. I really like the custom laminator: the motorized heated rollers apply just the right amount of heat and pressure, and allow the perfect amount of time to laminate the photo-resist film neatly onto the brass or steel. Also included is a tall etching tank with a clever clip support that suspends parts up to 3″ × 3″ inside the tank. An aquarium pump and aerator "bubbling nozzle" agitates the sodium hydroxide for better and faster etching. Although this kit costs over $100, I know I'll use it again and again.

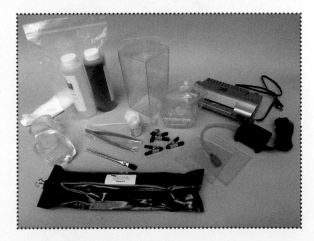

http://www.micromark.com/MICRO-MARK-PRO-ETCH-PHOTO-ETCH-SYSTEM,8346.html

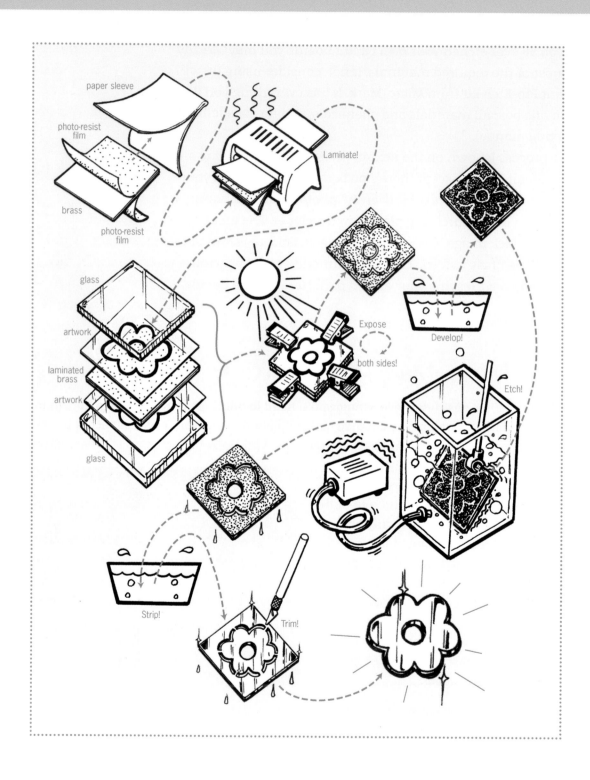

1. Create Your Image

Photo-etching can create very small details and features, down to a tiny 0.010″ or so. However, the smaller the feature, the more care you will have to take in keeping the image razor sharp and the exposure blur-free. Also, the finer the detail, the more vigilant you will need to be when developing and etching. Create your image as a bitmap in Photoshop or vector art in Illustrator, or even just go old school with hand-drawn images. Add dotted lines to aid bending or folding the finished shape. Connect all of the parts and shapes with thin lines so that when etched they stay together and don't fall off into the tank of acid. Leave a generous border of unetched brass around the parts to act as a frame to hold it all together while etching. Print out your image on clear film with dark black ink. The clear parts of the image will not etch—those are the resultant metal parts.

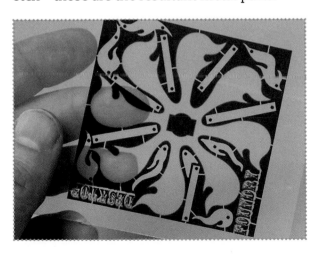

Unlike etching a PC board, in this process the acid will etch through *both sides* of the brass at the same time. You can choose to have a solid area of resist to protect the back, resulting in a simple part. You can also have slightly different images on the front and back. Where the areas of the images are the same, the brass is etched completely through, making a hole. Where the images are different, etching only happens halfway into the metal and only from one side, making an engraved feature like the tiny bolt heads or markings in the above examples.

2. Transfer to Metal

First, prepare the surface of the metal. It must be grease-free and perfectly clean. Scrub with a moistened polishing pad and handle the metal by the edges only—no greasy fingerprints! Peel off the protective carrier and adhere the photo-resist film to the metal with a little water. Smooth out any trapped air bubbles and excess water. Repeat for the other side. Then place the film-brass-film sandwich in a protective paper-cover sleeve and run it through the laminator. The heat and pressure adheres the photo film to the brass.

3. Expose to Light

Place the printed image over the film, clamp between plexiglass, and expose to UV light.

Use the bright noon sun for 15 seconds or a 100W bulb with a longer exposure time of up to 15 minutes or more. When the resist film is properly exposed, you should see a dark purple image, the negative "shadow" of your artwork. Repeat the exposure on the backside, in my case with no image (solid resist).

4. Strip

Remove the clear protective films from both sides and immerse the brass in a diluted bath of sodium hydroxide developer. The unexposed areas will soften and wash away, exposing the bare brass. Gently brush the surface to help remove the softened film until you have crisp edges on all the smallest features. Rinse clean with water.

5. Etch

Immerse the brass completely in a tank of ferric chloride. Use a support clip to periodically lift the part out of the tank to check on its progress. The etchant is nasty stuff, so wear protective gloves and goggles and work in a space with good ventilation. For best results also use an aerator in the tank—the gentle bubbling agitation will ensure a flow of fresh etchant.

Periodically lift the part from the solution and check the progress. Carefully brush away spent etchant over small details to speed up the process. Depending on the amount of metal to be etched and the level of detail, it may take 20 minutes or more to completely etch your design.

6. Clean, Trim, and Assemble

Rinse the finished part under warm water to remove all the etchant. Double-check the finest details and return it to the etching tank if it needs more time. When it's done, dip the finished part in a tray of undiluted sodium hydroxide to dissolve and remove the purple resist material.

When all the exposed brass is etched away and small details are clearly etched—it's done!

Gently brush away any stubborn resist, then rinse in clean water. Done!

Trim the parts from the supporting brass runners using a tiny snips or an X-Acto knife. Carefully burnish with a polishing pad for a shiny look.

I did a second etch to make the phoenix and some other trim pieces. I tapped the center with a center punch to make a slight dimple to act as a bearing.

Then I folded it to make the wings and head of the phoenix, and twisted the blades to make the turbine fan. When assembled the phoenix/fan rests on a sharpened rod and spins freely on the slightest bit of rising hot air. See it in action two ways!

Look in the corner of this book and use your thumb to flip the pages quickly: the turbine spins over the flame!

Scan this QR code with your smartphone to see a demo video of the spinning turbine.

Helpful Etching Tips and Tricks

KEEP IT SIMPLE!

On the fan blade artwork, I included some other finely detailed parts and filigreed lettering just to see how it would work. Bad idea—much of the detail "blew out" during developing and etching. Go for cool detail but don't make your parts any more complicated than you really need to.

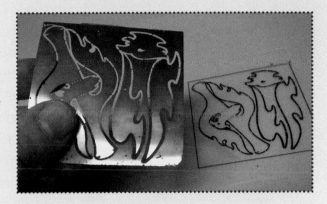

BE EFFICIENT!

My first design (the fan) had much more exposed area to be etched away than was really needed. That depletes the etchant solution faster and slows down the process. My second design (the phoenix) was much better: I surrounded the parts with plenty of extra brass. You really only need to etch away a thin outline around your shapes—not the entire sheet. Conserve the etchant and use that surrounding area to make other smaller shapes.

KEEP IT MOVING!

Check on your parts in the etching tank every couple of minutes. Rotate the orientation of the part often and place it in a different position in the tank to ensure the entire surface of the part is getting a good flow of fresh etchant.

Hack a Yakitori Grill

One of the most memorable and delicious aspects of travel is sampling the local foods. A trip to Japan gave me a chance to enjoy favorites like *takoyaki* (octopus fritters), *okonomiyaki* (cabbage frittatas), and other Asian eats in their native setting. A new (to me) treat was *yakitori*, a simple bar food snack of grilled chicken.

In the Tokyo neighborhood of Shinjuku, I saw (and smelled) enticing restaurants featuring sizzling streetside grills. Unlike big American grills that cook anything from burgers to ribs to steaks, these scaled-down grills are designed to do one thing, and one thing only: skewers. Short skewers loaded with chicken, asparagus, meatballs, and other simple ingredients spanned the narrow troughs of red-hot coals. The suspended foods cooked quickly and without burning or sticking to a grate or grill surface. And the offerings included nearly all the parts of the chicken, from succulent breast (*torinku*) to crunchy cartilage (*nankotsu*) and delicate, crispy chicken skin (*torikawa*). Yum!

Back home, I wanted a way to cook yakitori myself, so I came up with this easy-to-make grill design and some specially·designed roll-proof, double-crook skewers. Use them to try delicious yakitori recipes.

The basic design is simple: a single piece of sheet metal is rolled into a half cylinder to serve as the trough for the coals. The cake pans help form the trough. The grill stands on metal legs with insulated cork feet. Pop rivets hold it all together.

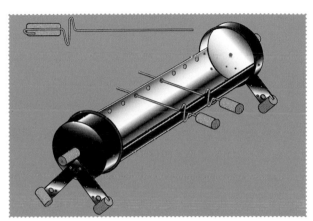

Maker Tip

CAUTION: Don't use galvanized steel for the main body of the grill. The zinc coating gives off dangerous fumes when heated—not good for a food-making project!

Materials and Tools

6"-diameter aluminum cake or deep-dish pizza pans
6" L-strap steel × 2
12" × 24" × 0.019" or thicker aluminum sheet
1/8" diameter stainless steel rod
Assorted 1/8" pop rivets and washers
Corks × 13 (use wine bottle corks made from *real cork*, not the new synthetic ones)
¼" D × 2" flat-head screws and nuts
High-temperature paint (used for grills or wood stoves)
Measuring tape
Straightedge
Hammer
Center punch
Pop rivet tool
Drill with ⁵⁄₃₂" and ³⁄₃₂" bits

1. Prepare the Sheet Metal

Sand any sharp edges or burrs on the aluminum sheet with 100-grit sandpaper. For extra safety, wear gloves when handling the sheet metal.

Measure and mark a series of 5/32″ holes that will serve as holders for the skewers. Make a line ½″ below the 24″ edge of the aluminum sheet. Starting 3 inches from the end, measure and mark the holes on 2″ centers. Use the center punch and hammer to make dimples to keep the drill bit from wandering when you drill. Then drill all 10 holes.

Drill 5 holes along each end of the sheet. These holes will be for connecting the pop rivets to the rims of the end caps. Draw a line 1″ from the edge, then measure, mark, and punch the position of a hole at the center, 6″ from each side. Then do the same for holes 1″ from each side, and holes 3½″ from each side.

2. Prepare the Pans

Find the center of the pan, then mark it and punch. Draw a line from the center to the rim of the pan.

Use this line to center the L-strap legs as shown, and carefully mark the location of the L-strap holes on the cap.

Use a block of wood as a support inside the pan. Center punch and drill all three holes. Do the same thing on the other pan so that they're identical.

Use pop rivets, with washers on the inside, to fasten the L-strap leg to the

outside of the pan. Repeat with the other pan and leg.

Gently roll the aluminum sheet to the 6″ diameter curve. Don't try to bend it all in one pass; just slightly curve it against a tabletop and slowly and gradually roll the sheet to bend the curve as you go. Check your progress using the pan.

Now "dry fit" the parts. Place the curved trough inside the pans. Do this on a flat, even surface; make sure all 4 feet are level and touching at the same time. Hold the pans tight against the curved edge of the sheet and mark the location of the holes in the sheet on the inside of the caps.

Maker Tip

I adjusted the position of the side with the holes so that it was ½″ lower than the other side—this will help level the double-crook skewers if you want to make them and use them with the grill. If you only want to use ordinary bamboo skewers, just center the trough. In any event, it's not critical.

Now use the holes already drilled in the trough to mark the locations of the matching holes in the end caps. Use a marker to carefully mark the position of each hole on the inside rim of each end cap.

Strike the marks gently with a center punch, then flip over and punch the same marks from the other side to convert them to dimples on the *outside* of the rim. That will make it easier to drill from the outside. Support with a wood block inside and drill the matching ⁵⁄₃₂″ rim holes.

3. Assemble and Paint

Line up the holes on the trough and connect it to the pans with pop rivets. Use washers to back up on the inside and to ensure snug riveting.

Mask off the inside surfaces with paper and masking tape. Then paint with two coats of high-temperature stove paint. Let dry overnight.

Drill a ⅛″ hole through the axis of 2 corks. Use a small flathead bolt to fasten the cork to the center hole on the end plate, with a washer and nut on the inside. Tighten the nut very snugly to pull the flathead flush with the cork—you don't want to touch the metal bolt when you pick up the grill by these insulated cork handles.

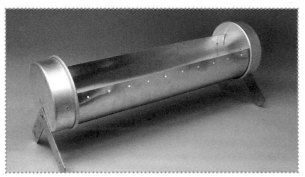

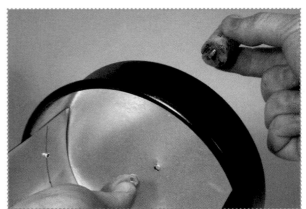

Use a hobby knife to make a slit along the length, but not all the way to the end of a cork, as shown. Carve out a little more cork material to make a slot that will fit snugly on the edge of the foot. Make three more cork pads, one for each foot. If the corks are loose, use some wire to poke through the cork and the hole in the leg. Twist the wire to secure.

Drill a ³⁄₃₂″ hole through the center axis of a wine cork and drill a second shallow ³⁄₃₂″ hole between the center and the edge of the cork.

Thread the cork on the short end of the skewer. Use the needle-nose pliers to make a very tight U-shape bend at the end of the wire.

4. Make the Skewers

To make the double-crook skewers, cut stainless steel rods to 14″ lengths. Use vise grips or pliers to make the bends shown. Dimensions aren't critical but make the bends into a zigzag shape.

Then slide the bent end into the second hole on the cork. This gives the skewer an insulated handle that won't spin. Make as many skewers as you like.

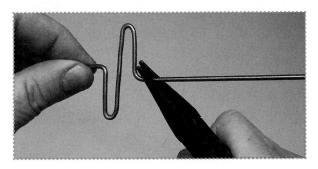

If you don't want to make these special skewers, you can use any kind of skewers. Look for *flat* cross-section bamboo skewers if you can find them—they'll rest on the grill edge without rolling when you flip them.

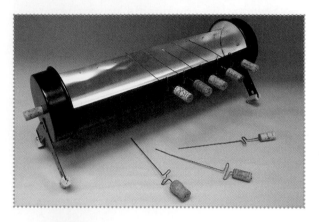

I salute *Make:* reader Greg Lehman, who sent me some great pictures of his own build of the Yakitori Grill in action. It has clever tangram-inspired graphics created by masking off shapes and lettering before painting. He also substituted square-shaped pans, which is a really great mod. Square pans may be easier to find and also make a slightly deeper charcoal tray for better heat.

Cook!

Put the grill on a fireproof surface away from any flammable vegetation or structures. Use crumpled paper and make a mound of charcoal in the center of the grill. For best flavor use *binchotan* (special Japanese high-carbon-content "white" charcoal) or mesquite, and avoid chemical starters. Light the coals and let them burn until uniformly covered in white ash—about 30 minutes. Can't wait? Use your Shop-Vac with the hose on the outlet as a blower to fan the flames. You'll have red-hot coals in just a few minutes, but be *very* careful—don't blow hot embers all over; go easy!

PHOTOS COURTESY GREG LEHMAN

Recipe for Ninja Chicken Sticks

(Sounds like more fun than yakitori chicken on skewers.)

1 lb boneless chicken thighs, with or without skin

¾ cup mirin (sweet rice wine)

½ cup soy sauce

½ cup sake

¼ cup sugar

Skewers

Combine mirin, soy sauce, sake, and sugar in a small pan and boil over medium heat until slightly thickened.

Cut chicken into bite-sized chunks or strips. Thread the meat evenly on the skewers, centering on the skewer to fit inside the grill. If you're using bamboo skewers, soak them in water first to prevent burning.

Spread the hot coals to make a uniform layer along the middle of the trough. Insert the skewers into the holes. Rotate the skewers every few minutes, brushing on more sauce. Repeat until golden brown.

You can "yaki" more ingredients besides chicken. Kushiyaki fare includes *ikada* (scallions), *butabara* (pork), *piman* (green pepper), *asuparabekon* (asparagus), and more.

Maker Tip

WARNING: Grills can be hot! And pop rivets can fail, and aluminum can degrade over time. Seriously, be careful, and discard the grill if any parts look as if they're losing their integrity.

Bonus Material

The bonus material included here will help you complete the projects in the book. I'll discuss some of the tools and materials used in the projects and point you to where you can find some of the less-common stuff. Then, I'll show you my solution to the electronics challenge in Chapter 2, as well as a reader's alternate solution. Finally, I'll provide you with clip-and-use versions of the graphics that go with some of the projects. Be sure to look at the makerfunbook.com website for more material and an additional project: "Industrial Design for Makers!" You can also download and print PDFs of the clip-and-use pages included here.

Tools, Materials, and Sources

Here are a few special tools and materials that I use in making toy prototypes.

Draw saw Sometimes called a Japanese saw, this handsaw has a very sharp blade, and teeth that are *backward* so that they cut on the pulling stroke instead of the pushing stroke. This trades off some power for precision and smooth cutting. The blade is loaded in tension, which keeps it straight and unbowed, making the cuts more controlled.

Plastic drill bits Ordinary drills are great for drilling thick wood or metal, but can be a problem when drilling holes in thin plastic like acrylics. The drill bit tends to catch suddenly just as it cuts through, causing the plastic to ride up the bit, often ruining the hole, or worse, chipping or breaking the plastic. If you'll be doing lots of projects using plastics, get a few special plastic drill bits. They have a sharper tip angle and ground sides that won't grab.

Surface gauge This tool is used by model and pattern makers, but you'll find it handy for lots of things. It's essentially an adjustable pointer on a base. Working on your flat tabletop, place the gauge next to a feature on an object you want to measure and adjust the thumbwheels to set the pointer to that location. Then as you slide the gauge along the tabletop, the pointer will indicate the exact same height anywhere on the surface of the object. I find it perfect for marking and scoring trim lines on vacuum formed parts. (See the Chapter 4 project, "Kitchen Floor Vacuum Former.")

Compass tool This handy version of a compass has a split hub so it can hold just about anything as one of its legs: pencil, scribe, fat permanent marker, bow pen, brush, you name it. I've used it countless times to draw circles and mark arcs.

Dremel With the huge assortment of attachments you'll find many uses for a Dremel rotary tool: sanding, drilling, cutting, buffing, polishing, grinding, and wire brushing. Adjust the variable speed to match the job.

Hobby knife No project gets done in my shop without a trusty X-Acto™ hobby knife and a fresh #11 blade. Buy the blades by the box!

Sculpting tools The small angled hooks and special shapes are very handy for sculpting small details in clay or Sculpey™.

Scribe and solvent glue bottle Styrene sheet is super easy to work with. You can create shapes with it by scoring a line with a scribe tool and then bending and snapping the plastic apart. Styrene also drills and sands easily. Attach pieces together instantly by solvent-bonding using MEK applied with a brush, or more easily with a solvent bottle with a capillary tube tip. Find styrene and acrylic sheet in various thicknesses at your local plastic supply store. I go to TAP Plastics (www.tapplastics.com/). If you don't want to buy a full 4′ × 8′ sheet, you may be able to find some useful pieces in their cut-off and remainder bin—just ask. They'll also have solvent adhesives, plastic drills, heating strips, scoring tools, polishes, casting and mold-making materials, and more.

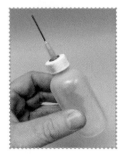

Vacuum former The commercially produced Nichols Therm-O-Vac has served me well and sees lots of use. It comes with adapter plates for using any size sheet of plastic, and folds up into a compact case with a carrying handle. (Or follow the instructions in Chapter 4 to build your own inexpensive vacuum former.)

Scan the QR code and watch the Nichols Therm-O-Vac make a guest appearance in the movie *F/X*.

I bought mine through IASCO-TESCO, an industrial arts supply catalog.

http://iasco-tesco.com/catalog/

IASCO-TESCO is also a great source for lots of industrial arts stuff: plastisol, styrene sheet, injection molding pellets, casting and mold-making materials, and much more.

Chapter 2: Easy Electronics
LED Challenge

Here's one solution to the wiring logic challenge. It will give outcomes of 1, 2, 3, 4, 5, or 6 LEDs lighting up. This design will yield the same average roll of a real die, but it won't be statistically "fair."

A real die has an average roll of 3.5:

(6 + 5 + 4 + 3 + 2 + 1) / 6 total number of possible outcomes =

(21) / 6 = 3.5

Here's the table of switch inputs and LED outputs. The first column is all of the possible states in which the ball switches can land. For example, LLL means the 4's ball has landed in the left side position, the 2's ball has landed in the left side position, and the 1's ball has landed in the left side position. There are 8 possible outcomes in all. This is binary coded decimal: 1 means on, and 0 means off. The last column is the total number of LEDs that are lit for each state.

	4	2	1		
LLL	1	1	0	=	6
LLR	1	1	0	=	6
LRL	1	0	0	=	5
LRR	1	0	0	=	4
RRR	0	0	1	=	1
RRL	0	0	1	=	1
RLL	0	1	1	=	3
RLR	0	1	0	+	2
Total					28

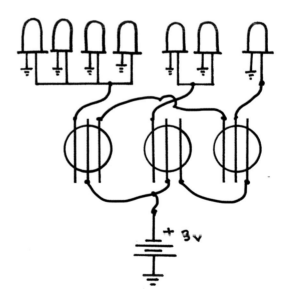

Now, 28 divided by 8 possible states yields an average of 3.5, the same as a real die, but it doesn't roll with the same even probabilities for all outcomes—there are twice as many 6s and 1s as there are 2s, 3s, 4s and 5s.

A salute to *MAKE:* reader and engineer Richard Ake, who sent me his clever solution to this LED die challenge.

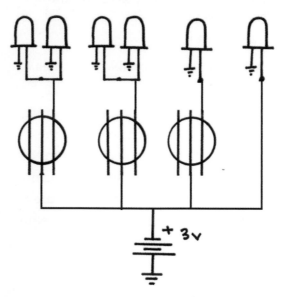

He describes it as "Simpler, but no more fair . . . Plus I save one LED so [it's] cheaper."

Here's another version.

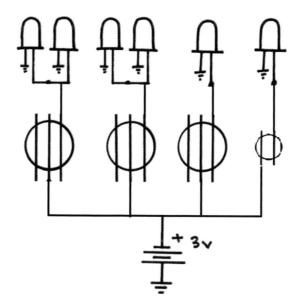

Turn it off by laying it on its side.

The roll table is 1, 2, 3, 3, 4, 4, 5, 6, so while the original is weighted to the extremities, mine is weighted to the center.

Richard says, "A 3-position switch and a 2-position switch would result in 6 answers, but the wiring of the LEDs would be challenging. I'm considering that. I think the solution is to use the LEDs not only as lamps but to also take advantage of the diode feature."

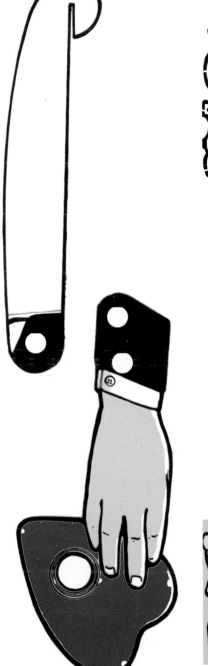

PSI
pixels seen in

NightVison

Twilight
Resolution

Keypad template—full size

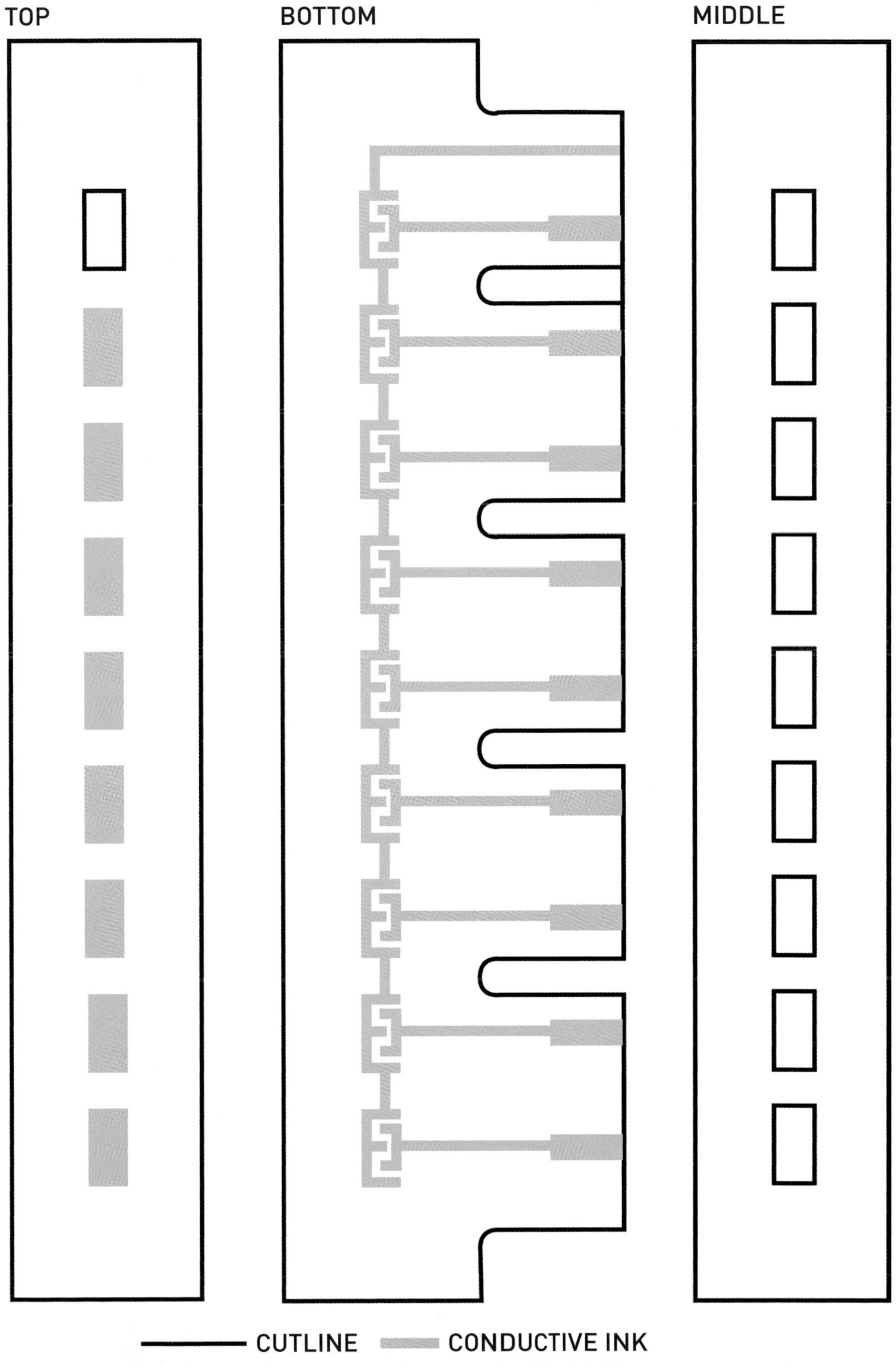

TOP   BOTTOM   MIDDLE

CUTLINE   CONDUCTIVE INK

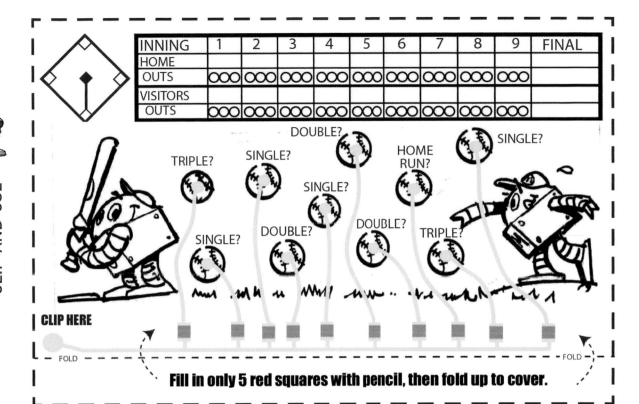

INNING	1	2	3	4	5	6	7	8	9	FINAL
HOME										
OUTS	○○○	○○○	○○○	○○○	○○○	○○○	○○○	○○○	○○○	
VISITORS										
OUTS	○○○	○○○	○○○	○○○	○○○	○○○	○○○	○○○	○○○	

TRIPLE? SINGLE? DOUBLE? HOME RUN? SINGLE?

SINGLE?

SINGLE? DOUBLE? DOUBLE? TRIPLE?

CLIP HERE

FOLD   FOLD

Fill in only 5 red squares with pencil, then fold up to cover.

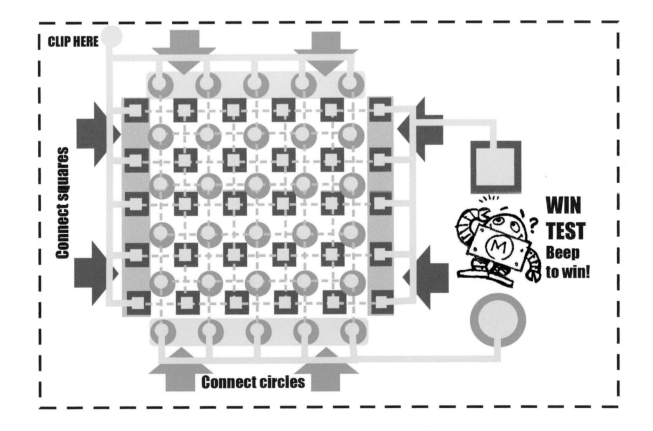

CLIP HERE

Connect squares

Connect circles

WIN TEST Beep to win!

CLIP HERE

1

2

START

3

CLIP HERE

tape
penny
here

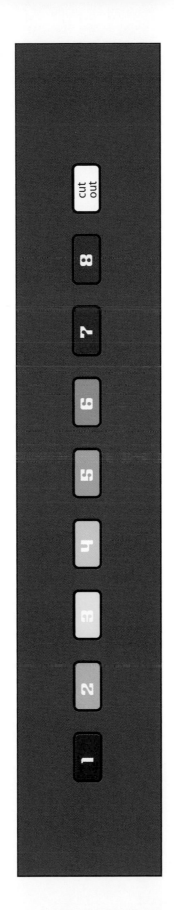

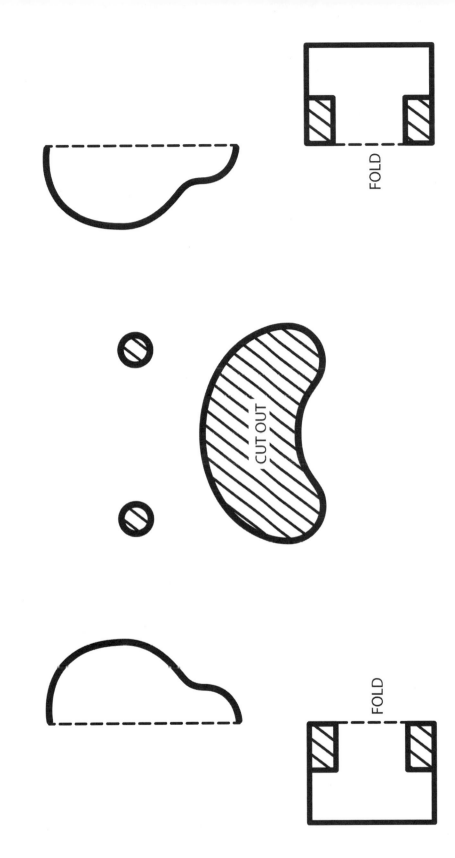

FOLD

CUT OUT

FOLD

Mad Monster Candy Snatch game

face label

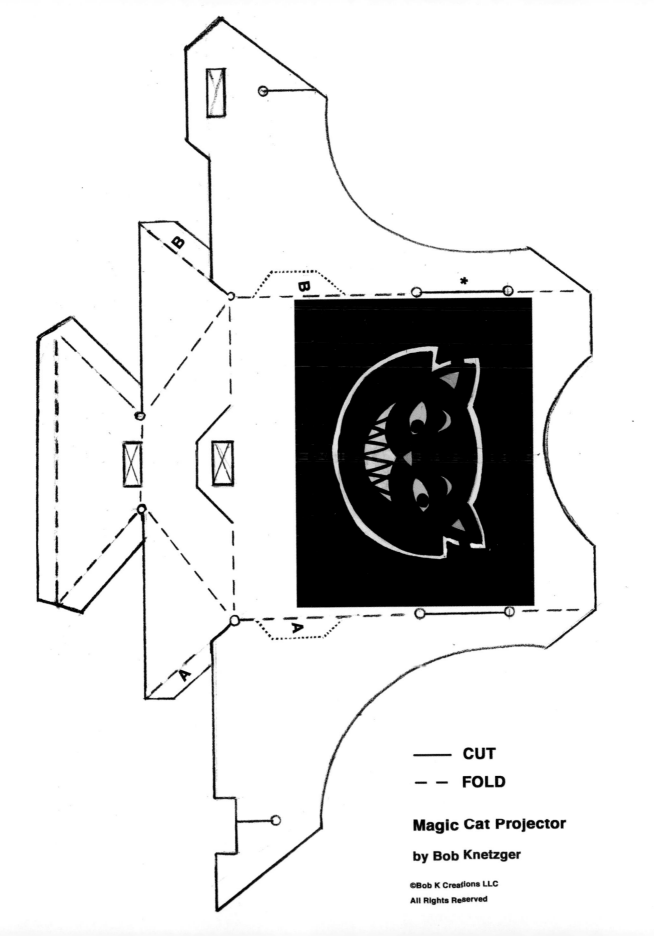

CUT

FOLD

Magic Cat Projector

by Bob Knetzger

©Bob K Creations LLC

All Rights Reserved

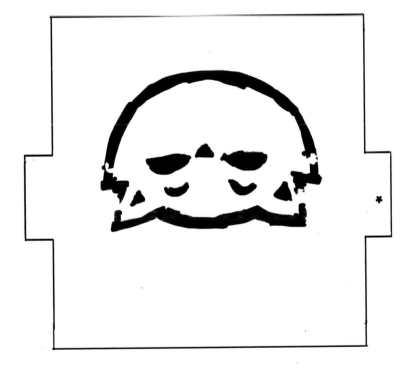

———— **CUT**

– – **FOLD**

Magic Cat Projector

by Bob Knetzger

©Bob K Creations LLC
All Rights Reserved

Index